ACTA PHILOSOPHICA FENNICA

Vol. 91

LOGIC, LANGUAGE AND GAMES

GABRIEL SANDU

HELSINKI 2015

VERTAISARVIOITU
KOLLEGIALT GRANSKAD
PEER-REVIEWED
www.tsv.fi/tunnus

ISBN 978-951-9264-81-3
ISSN 0355-1792

Hakapaino Oy
Helsinki 2015

Contents

Introduction

The present study is based on existing publications and lecture notes organized in a new way. It centers around game-theoretical semantics for first-order logic and its extension, Independence-friendly logic (IF logic). IF logic and its latest development, probabilistic IF logic rely heavily on game-theoretical results (von Neumann's minimax theorem). It is tailored made to model various game-theoretical phenomena like indeterminacy and signaling. For this reason a great part of the book contains introductory material to game theory and some of its solution concepts, which lead, in the end, to (Nash) equilibrium semantics.

Chapter 1

Dependence and independence of quantifiers

1.1 Quantifiers

According to a common assumption which goes back to Frege, quantifiers receive their meaning from "ranging over" a class of entities. An existential quantifier $\exists x$ prefixed to a formula φ with a single free variable x says that the class of individuals satisfying φ, denoted $\|\varphi\|$, is not empty. A universal quantifier $\forall x$ prefixed to φ says that $\|\varphi\|$ is the entire universe. Thus it is natural, following Frege, to take the semantic value of a quantifier to be a second-order predicate, that is, a collection of subsets of the universe. Given a model M, the existential quantifier is identified with the collection of all nonempty subsets of M, and the universal quantifier is identified with the collection whose only member is the universe itself:

$$\begin{aligned} \exists_M &= \{X \subseteq M : X \neq \varnothing\}, \\ \forall_M &= \{M\}. \end{aligned}$$

The truth-conditions of existential and universal formulas fall out quite naturally from these semantic conventions:

- $\exists x \varphi$ is true in M if and only if $\|\varphi\| \in \exists_M$, and
- $\forall x \varphi$ is true in M if and only if $\|\varphi\| \in \forall_M$.

This approach can be generalized quite easily. Mostowski has enlarged the syntax of first order logic to cover quantifiers other than $\forall x$ and $\exists x$. He realized that every monadic second-order relation (i.e., relation which has only one argument for sets) can be represented by a quantifier Qx which applies to a formula φ. In this way we obtain extensions of first-order logic with new formulas of the form $Qx\varphi$. Examples of generalized quantifiers include:

$$(\exists_{\geq n})_M = \{X \subseteq M : |X| \geq n\},$$
$$(Q^R)_M = \{X \subseteq M : |X| > |M - X|\},$$
$$(Q_\infty)_M = \{X \subseteq M : X \text{ is infinite.}\}$$

Here $(\exists_{\geq n})$ stands for "there exist at least n individuals," (Q^R) stands for "more individuals than not," and (Q_∞) stands for "there exist infinitely many individuals."

This way of looking at quantifiers has turned out to be very fruitful for the study of the semantics of natural language. The main focus of the present book is, however, on a different topic, to which we now turn.

1.2 Dependence of quantifiers

Although the "ranging over" is an important part of the semantic job of quantifiers, it is not the only one. One of the revolutionary aspects of modern logic consists in considering valid inferences that involve multiple quantification. When at least two quantifiers, say $\forall x$ and $\exists y$, occur in a formula we can express that one of them lies within the scope of the other by simply writing one after the other $\forall x \exists y$. The intriguing question is how does the "ranging over" paradigm fit with the dependence of quantifiers? The former requires considering one quantifier at a time, while the latter seems to demand a mechanism that "glues" these successive steps together.

The linear ordering of the quantifiers in the Frege-Tarski tradition was felt to be the natural one because it allowed giving a relatively straightforward answer to the question. Given a formula in which the quantifiers are linearly ordered, one can consider one *singly* quantified formula at a time and still account for the dependence of a quantifier on the previous one in the sequence. Consider a simple sequence of quantifiers as in the scheme

(1.1) $$Q_1 x_1 Q_2 x_2 R(x_1, x_2)$$

where each Q_i is either $\exists$ or $\forall$. In the Frege-Tarski tradition one defines the truth-conditions of such a formula in terms of a series of one-place relations. The "gluing together" is achieved by a relative expression synonymous with "for each one of which." Thus the interpretation of (1.1) is specified in the following steps:

> (1.1) is true iff there are q_1 a's for each one of which there are q_2 b's such that $R(a, b)$.

Here q_1 and q_2 are the semantic conditions associated with the quantifiers Q_1, and Q_2, respectively. (It is not easy to give an English paraphrase of the intended interpretation. The paraphrase we have chosen fits best the existential and cardinal quantifiers.)

1.3 Independence of quantifiers (Sher)

Against the backdrop of the dependence between quantifiers sketched above, there are several ways to conceive of independence. We shall consider two such notions here, which have a prominent place in the literature. One of them, due to Sher (1990), construes an independent quantifier as a syntactically non-nested quantifier corresponding, semantically, to a non-relativized, separate domain. It does not lead to an increase in expressive power. The second notion, called the *cluster domain* or *massive nucleus interpretation*, emerged from the work of Henkin (1961) and Barwise (1979). It typically increases the expressive power of a language beyond that of the language involving the individual quantifiers.

If Q_1x_1 and Q_2x_2 are two generalized (monadic) quantifiers, then their independence, denoted by

$$\begin{pmatrix} Q_1x_1 \\ Q_2x_2 \end{pmatrix} R(x_1, x_2) \tag{1.2}$$

amounts to saying that the projection of R^M onto its first coordinate belongs to Q_1 and the projection of R^M onto its second coordinate belongs to Q_2:

$$\{a : \exists b R(a,b)\} \in (Q_1)_M$$

$$\{b : \exists a R(a,b)\} \in (Q_2)_M$$

This is just another way of saying that (1.2) is true if and only if:

(i) there are q_1 a's for each one of which there is at least one b such that $R(a, b)$ is the case, and

(ii) there are q_2 b's for each one of which there is at least one a such that $R(a, b)$ is the case.

Obviously the logical force of (1.1) is distinct from that of (1.2).

Here are a couple of examples from natural language which motivate Sher's notion of independence:

(1.3) Three frightened elephants were chased by a dozen hunters.

(1.4) Nobody loves nobody (anybody).

An independent reading of (1.3), represented syntactically by

$$\begin{pmatrix} 3x_1 \\ 12x_2 \end{pmatrix} (E(x_1) \wedge H(x_2) \wedge C(x_1, x_2)) \tag{1.5}$$

simply says that the set of hunted frigthened elephants

$$\{a : \exists b(E(a) \wedge H(b) \wedge C(a, b))\}$$

contains three members, whereas the set of frightened-elephant hunters

$$\{b : \exists a(E(a) \wedge H(b) \wedge C(a, b))\}$$

contains twelve members.

On the other hand, the nested, dependent reading

$$3x_1 12x_2(E(x_1) \wedge H(x_2) \wedge C(x_1, x_2)) \tag{1.6}$$

has it that each frightened elephant (of which there are three) is such that twelve hunters are after it.

Sentence (1.4) is due to Robert May. If it were to be represented by the nested

$$\neg\exists x_1 \neg\exists x_2 L(x_1, x_2) \tag{1.7}$$

then it would be equivalent to

$$\forall x_1 \exists x_2 L(x_1, x_2)$$

which says that everybody loves someone.

If, on the other hand, the two quantifiers were taken to be independent,

$$\begin{pmatrix} \neg\exists x_1 \\ \neg\exists x_2 \end{pmatrix} L(x_1, x_2) \tag{1.8}$$

whose truth-conditions would be expressed by

$$\neg\exists x_1 \exists x_2 L(x_1, x_2) \wedge \neg\exists x_2 \exists x_1 L(x_1, x_2)$$

which says that the relation L^M is empty. This seems to be the intended interpretation of (1.4).

The independence of two quantifiers $Q_1 x_1$ and $Q_2 x_2$ does not increase expressive power because

$$\begin{pmatrix} Q_1 x_1 \\ Q_2 x_2 \end{pmatrix} R(x_1, x_2) \Longleftrightarrow Q_1 x_1 \exists x_2 R(x_1, x_2) \wedge Q_2 x_2 \exists x_1 R(x_1, x_2) \tag{1.9}$$

and the same goes for the general case:

$$\begin{pmatrix} Q_1 x_1 \\ \vdots \\ Q_n x_n \end{pmatrix} R(x_1, \ldots, x_n) \Longleftrightarrow$$

$$\begin{aligned} & Q_1 x_1 \exists x_2 \ldots \exists x_n R(x_1, \ldots, x_n) \\ \wedge\ & Q_2 x_2 \exists x_1 \exists x_3 \ldots \exists x_n R(x_1, \ldots, x_n) \\ & \vdots \\ \wedge\ & Q_n x_n \exists x_1 \exists x_2 \ldots \exists x_{n-1} R(x_1, \ldots, x_n). \end{aligned}$$

In particular, when $Q_1, \ldots, Q_n$ are the standard first-order quantifiers, then

$$\begin{pmatrix} Q_1 x_1 \\ \vdots \\ Q_n x_n \end{pmatrix} R(x_1, ..., x_n)$$

expresses a first-order condition.

Sher (1990) generalizes this kind of independence in several ways. The one that is relevant for our present concerns is that which considers independent blocks of quantifiers, like

$$\begin{pmatrix} \forall x \exists y \\ \forall z \exists u \end{pmatrix} \varphi(x, y, z, u)$$

The basic idea is the same as before, i.e., the first two domains of the relation satisfy the condition corresponding to the first block of quantifiers, and the last two satisfy the condition corresponding to the second block:

$$(1.10) \quad \begin{pmatrix} \forall x \exists y \\ \forall z \exists u \end{pmatrix} \varphi(x, y, z, u) \Longleftrightarrow$$
$$\forall x \exists y \exists z \exists u \varphi(x, y, z, u) \wedge \forall z \exists u \exists x \exists y \varphi(x, y, z, u)$$

The analogy with (1.2) should be obvious. We shall consider some examples after reviewing another notion of independence.

1.4 Independence of quantifiers (Henkin, Barwise)

In this section, we shall take a look at a notion of independence due to Barwise (1979). Unlike Sher's notion, which considers domains in isolation, Barwise's analysis of independent (branching) quantifiers forces us to consider a cluster of domains simultaneously, in their mutual relationships. This notion is formulated for monotone increasing (*most*) or monotone decreasing (*few*) quantifiers, for which it yields essentially a second-order property.

A (monadic) quantifier Q_1 is said to be *monotone increasing* if, whenever we have $Q_1 x \varphi(x)$ and $\forall x(\varphi(x) \rightarrow \psi(x))$, we also have $Q_1 x \psi(x)$. The notion of a *monotone decreasing* quantifier is defined analogously. For two monotone increasing quantifiers $Q_1 x$ and $Q_2 y$, Barwise's definition of branching quantification is

$$\begin{pmatrix} Q_1 x \\ Q_2 y \end{pmatrix} R(x, y) \Longleftrightarrow$$
$$\exists X \exists Y [Q_1 x X(x) \wedge Q_2 y Y(y) \wedge \forall x \forall y (X(x) \wedge Y(y) \rightarrow R(x, y)]$$

This definition generalizes to independent blocks of quantifiers, like

$$(1.11) \quad \begin{pmatrix} \forall x \exists y \\ \forall z \exists v \end{pmatrix} V(x, y, z, u) \Longleftrightarrow$$
$$\exists R \exists S [\forall x \exists y R(x, y) \wedge \forall z \exists u S(z, u)$$
$$\wedge \forall x \forall y \forall z \forall u [R(x, y) \wedge S(z, u) \rightarrow V(x, y, z, u)]]$$

Obviously, Barwise's notion of independence is different from Sher's as may be seen in connection with the so-called Hintikka sentence:

(1.12) Some relative of each villager and some relative of each townsman hate each other.

If (1.12) were to represent the kind of independence illustrated by (1.10), then it would have the logical force of

$$\forall x \exists y \exists z \exists u [V(x) \to R(x,y) \land T(z) \land R(z,u) \land H(y,u)] \\ \land \forall z \exists u \exists x \exists y [T(x) \to R(z,u) \land V(x) \land R(x,y) \land H(x,y)]$$

which says that for each villager there is a relative which is in a hate relation with a relative of some townsman, and the same holds reciprocally for each townsman. This account is not, however, the one favoured by Hintikka. In order to obtain Hintikka's favourite interpretation, we have to go for the kind of independence expressed by (1.11):

> There are two sets, one containing at least one relative of each villager and the other containing a relative of each townsmen, such that each villager-relative in the first set hates all the townsman-relatives in the second, and *vice versa*.

The branching interpretation expresses essentially a second-order property.

The question whether Hintikka's sentence (1.12) has one or the other of these two interpretations gave rise to a lively exchange between Hintikka and Fauconnier. Fauconnier (1975) produced an example in which the massive nucleus interpretation is not always enforced:

(1.13) Some player of every football team is in love with some dancer of every football company.

Under the presupposition that men are in love with one woman at a time, the massive nucleum interpretation cannot be the intended one. The intended interpretation is captured in this case rather by Sher's notion of independence (1.10). Fauconnier concludes that the branching, independence interpretation does not apply to all quantified sentences in English and that the nestings of the four quantifiers is sometimes branching and sometimes linear.

We shall not enter into this debate in the present book. In any case, it is taken for granted that what distinguishes the second-order interpretation of the Hintikka sentence from its first-order interpretation is the massive nucleus interpretation. Sher (1990), and Westerstahl (1987) have developed accounts of branching quantification which generalize Barwise's definition to quantifiers other than the monotonically increasing ones.

The conception of independence which guides the present work is akin to the massive nucleum interpretation, except that it replaces second-order relations with functions. That replacement opens the door to the notions of functional dependence and independence of quantifiers which in turn are grounded in the game-theoretical account of quantifiers emerging from the work of Henkin and Hintikka. This account was devised for the standard quantifiers but it has interesting generalizations to connectives and other logical operators that behave like them.

1.5 Game-theoretical interpretation of dependence and independence

1.5.1 Semantic games for classical logic

On Henkin's account, the interpretation of a sequence of standard quantifiers can be displayed in terms of the strategic interaction of two players in a game:

> Imagine, for instance, a "game" in which a First Player and a Second Player alternate in choosing an element from a set I; the infinite sequence generated by this alternation of choices then determines the winner. If we let π denote the class of all those sequences for which the First Player is the winner, then the formula $[\exists v_1 \forall v_2 \exists v_3 \forall v_4 \ldots (\pi v_1 v_2 v_3 \ldots)]$ simply expresses the fact that the First Player has a winning strategy. (Henkin 1961, p. 179)

Following Henkin, we shall take the following steps:

a) We first give quantifiers a game-theoretical content. The universal and the existential quantifiers become moves that prompt the players to choose individuals from the relevant universe of discourse. $\exists x\varphi$ prompts the existential player (*Eloise*) to find an individual a that satisfies φ, and $\forall x\varphi$ prompts the universal player (*Abelard*) to produce an individual b that falsifies φ. This mode of thinking extends naturally to the logical connectives. For $(\varphi \vee \psi)$, the existential player will have to choose a disjunct, and for $(\varphi \wedge \psi)$, the universal player will have to choose a conjunct. In this new setting, the game associated with the sentence $\exists x \forall y R(x, y)$ gives rise to a sequence of choices: a choice of an individual a by the existential player followed by a choice of an individual b by the universal player. If a and b satisfy $R(x, y)$, then the existential player wins; if not, the universal player wins.

b) Once the individuation and the order of choices are established, the scopal dependences of quantifiers in a formula can be seen to express the knowledge of the players in the relevant games, under the tacit assumption that "being in the scope of a quantifier" is "knowing the move it prompted." In classical game theory such knowledge is encoded in the information set of a player for a particular position. This notion finds its way into the definition of a player's strategy, which is a function defined on all appropriate information sets in the game, and whose value indicates how the player should move in each position. Thus a strategy for the existential player in the game $\exists x \forall y R(x, y)$ is a 0-ary function that picks an individual a from the universe M. It is a winning strategy if for every $b \in M$ the pair (a, b) satisfies $R(x, y)$. In contrast, a strategy for the universal player is a unary function $f \colon M \to M$, which is winni! ng if the pair $(a, f(a))$ always falsifies $R(x, y)$.

c) Finally, we identify the truth or falsity of a sentence with the existence of a winning strategy for the existential or universal player, respectively, in the appropriate game. When the sentence the game is associated with is in prenex normal form, the functions that are the winning strategy of the existential player (when they exist) are known as Skolem functions; and the functions that are the winning strategy of the universal player (when they exist) are known as Kreisel's counterexamples.

Thus the game-theoretical setting brings in a correlation between (material) truth (falsity), strategies in extensive games and Skolem functions (Kreisel's counterexamples). It construes the dependence between quantifiers as informational dependence in semantic games.

This way of looking at a sequence of interdependent quantifiers is quite natural when one looks at the mathematical vernacular. The definition of standard of continuity is a standard example in the literature:

> A function f is *continuous* if for all x in the domain of the function and for all $\varepsilon > 0$ there exists a $\delta > 0$ such that for all y in the domain,
>
> $$|x - y| < \delta \quad \text{implies} \quad |f(x) - f(y)| < \varepsilon.$$

Following (a), "for all" receives the game-theoretical content of the universal quantifier, and "there exists" that of the existential quantifier. The scopal dependences between the four quantifiers in the definition of continuity are displayed by the first-order formula:

$$\forall x(\forall\varepsilon > 0)(\exists\delta > 0)\forall y(|x - y| < \delta \to |f(x) - f(y)| < \varepsilon) \tag{1.14}$$

Following (b), the dependence of $(\exists\delta > 0)$ on the first two universal quantifiers is indicated by the arguments of the existential player's strategy function in the appropriate game: (1.14) is true if and only if there is a Skolem function g such that for every $\varepsilon > 0$ and for all x and y we have $g(x, \varepsilon) > 0$ and

$$|x - y| < g(x, \varepsilon) \to |f(x) - f(y)| < \varepsilon.$$

Symmetrically, the dependence of $\forall y$ on the precedent quantifiers are spelled out in the falsity condition of (1.14):

$$\exists x(\exists\varepsilon_x > 0)\exists h_{x,\varepsilon_x}(\forall\delta > 0)(|x - h_{x,\varepsilon_x}(\delta)| < \delta \wedge |f(x) - f(h_{x,\varepsilon_x}(\delta))| \geq \varepsilon_x)$$

where x, ε_x and h_{x,ε_x} are Kreisel's counterexamples. Notice how in $h_{x,\varepsilon_x}(\delta)$ the universal player's prior choices act like parameters, while his opponent's (single) choice acts like a variable. More on this topic will be said in Chapter 3 and Chapter 4.

1.5.2 Game-theoretical account of independence

Once the game-theoretical interpretation of quantifiers is adopted, it makes sense to talk not only about a quantifier being informationally dependent on, but also it being informationally independent of other quantifiers. Henkin (1961) was the first one to provide examples and to develop this interpretation. He introduced branching quantifiers

$$\begin{pmatrix} \forall x \exists y \\ \forall z \exists u \end{pmatrix} R(x, z, y, u) \tag{1.15}$$

and interpreted them via the systematic use of Skolem functions. The sentence (1.15) is true if and only if there exist functions f and g such that

$$\forall x \forall z R(x, z, f(x), g(z)).$$

The Skolem functions f and g can be thought of as strategy functions in a game of imperfect information: Eloise must choose the value of y knowing only the value of x, and she must choose the value of u knowing only the value of z.

In Hintikka and Sandu (1989), languages for imperfect information, or independence-friendly languages (IF languages) were introduced with the purpose of having a general framework for expressing in linear notation quantifier dependences and independences of the sort considered by Henkin. These languages, described in detail in Chapter 3, contain quantifiers of the form

$$(\exists x/W)\varphi, \ (\forall y/W)\varphi,$$

where W is a finite set of variables. In this new notation (1.15) would be expressed by

$$\forall x \forall z (\exists y/\{z\})(\exists u/\{x, y\}) R(x, z, y, u) \tag{1.16}$$

As in the case of ordinary first-order languages, the new syntax is meant to display explicitly the knowledge patterns of the players in the appropriate games.

1.5.3 Independence as uniformity in extensive games

From a semantic point of view, IF languages illustrate nicely the notion of uniformity. Let us look at a few examples.

Consider the simplest pattern of independent quantifiers, as in the sentence

$$\forall x (\exists y/\{x\}) F(x, y)$$

Game theoretically it expresses the fact that the choice of a witness for the existential quantifier is independent of (does not depend on) the choice of a value for the universal quantifier. On the account of independence adopted here, this amounts to the requirement that any strategy function f of the existential player be constant. But then the existence of a winning strategy in the game comes down to the existence a single individual c such that for all a, $F(a, c)$. As a result, the above sentence is truth equivalent to $\exists y \forall x F(x, y)$.

The example illustrates a pattern of switching quantifiers, $\forall\exists \to \exists\forall$, which is well known in mathematics. For instance, one makes a distinction between a distribution on a test (every student has received a score), and a uniform distribution (there is a score that every student has received).

Our earlier definition of continuity provides another example of independence which involves four quantifiers. We notice that in this definition δ depends on both ε and the point x. A continuous function for which δ depends only on ε (and not on x) is called *uniformly continuous*. The difference between continuity and uniformly continuity comes down to the switching of quantifiers $\forall\forall\exists\forall \to \forall\exists\forall\forall$, which, in the setting of the present book, amounts to the difference between $\forall x\forall y\exists z\forall w$ and $\forall x\forall y(\exists z/\{x\})\forall w$. In the latter game Eloise's strategy f is uniform in x: for any two sequences (a, b), (c, d) of elements in the relevant universe, $f(a, b) = f(c, d)$, whenever $a = c$. This has the effect that (the values of) z depend only on (the values of) y.

In both examples, independence qua uniformity amounts to the switching of quantifiers. This might give the impression that quantifier independence qua uniformity is just another name for quantifier switching. But this is not so as the following variations on the pattern of four quantifiers show:

$$\forall x\forall y(\exists z/\{x\})(\exists w/\{y,z\})\varphi \quad \text{and} \quad \exists w\forall x\exists y(\exists z/\{x\})\psi$$

In the formula on the left, we want the first existential quantifier to depend only on $\forall y$, and the second to depend only on $\forall x$. But no switching of quantifiers will have that effect. Similarly, in the formula on the right we want the last existential quantifier to depend only on all the other existential quantifiers. Again, no quantifiers switching will do the job. Yet it might be possible to find ordinary (first-order) formulas which are equivalent with the given ones. But this is not always true. It will be seen that for certain choices of φ and ψ, the two formulas above express properties which are not first-order definable.

1.5.4 Independence as Nash equilibria in strategic games

There is another way to look at the independence between quantifiers in a game-theoretical setting. We introduce it by way of an example. On the conception of independence as uniformity both games

$$\forall x(\exists y/\{x\})x = y$$
$$\forall x(\exists y/\{x\})x \neq y$$

are nondetermined in every domain which contains at least two individuals. For in both games, the uniformity of the strategy functions for the existential player reduces them to constant functions. The result is that the first sentence is true iff there is an individual y which is equal to any x chosen by her opponent. But this is impossible. And it is false if the universal player can choose an element x which is distinct from any element y chosen by his opponent. But this is impossible too. And a similar reasoning applies to the second example.

From the vantage point of the uniformity interpretation, the two formulas are indistinguishable on any universe which has at least two individuals.

There is nevertheless a difference between the two examples that has been pointed out by Ajtai. Let us think of the two players choosing simultaneously elements of the relevant domain and of their odds to win it. The odds of the existential player of winning the second game must increase, and her odds of winning the first game must decrease, as the universe increases. That is, if the universe consists of three individuals, Eloise wins the first game with a probability of $\frac{1}{3}$ and the second game with a probability of $\frac{2}{3}$. And if the universe consists of four individuals, then the respective probabilities are $\frac{1}{4}$ and $\frac{3}{4}$. In order to account for such facts, we shall consider independence in the context of strategic games with mixed strategies and make appeal to solution concepts in game theory (Nash equilibria).

Chapter 2

Introduction to game theory

2.1 Extensive games with perfect information

Game theory consists, among other things, of a collection of hypotheses formulated in a precise way using mathematical tools. The hypotheses give rise to mathematical models called *games*.

Any game involves, roughly:

- a list of players

- a collection of possible choices for each player

- a function which specifies the payoffs of the players for each possible outcome.

When games are presented in extensive form, there is a sequential order of the moves which makes the interactions of the players in time quite explicit. We start our presentation with such games.

An extensive game of perfect information has the structure $G = (N, H, Z, P, (u_p)_{p \in N})$ where

- N is a finite set of players,

- H is the set of histories of the game,

- Z is the set of maximal histories (plays),

- $P : H - Z \to N$ is the player function which tells whose player's turn is to move, and finally

- u_p: $Z \to \{0, 1\}$ is the payoff function for player p.

An extensive game is often represented in the form of a tree, *Kuhn's tree*, as in the figure 2.1.

The root of the tree represents the initial history and the labelling attached to it marks the player who has the first move. Each of the three branches which

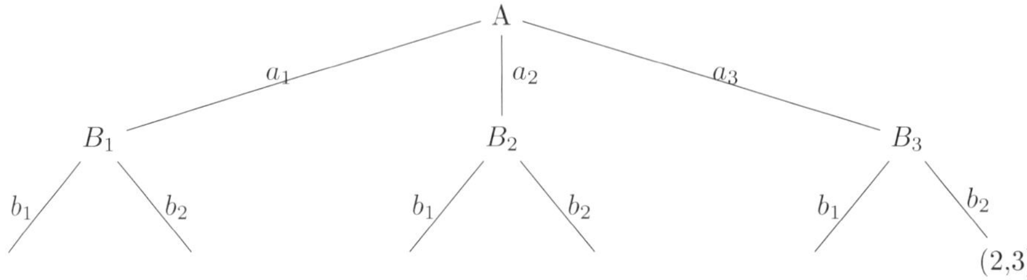

Figure 2.1: A Kuhn tree for a game

emerges from the root represents a possible choice (action) of the relevant player and leads to the next node (position) in the tree. Each such node prompts a move by one of the players such that each of the player's actions will generate further branches in the tree. Each maximal history represents a possible play of the game. The extensive game that is represented by the tree in the figure has a special property: At each node (position), the player who makes the next move knows what has happened earlier in the play. That is, at node B_1 the player B knows that player A has chosen a_1; at node B_2 he knows that the A has chosen a_2 and at node B_3 he knows that A has chosen a_3. For this reason such an extensive game is called a game with *perfect information.* Notice also that each maximal history in Z comes with a sequence of payoffs, one for each player. Thus $u_A(a_3, b_2) = 2$ and $u_B(a_3, b_2) = 3$.

2.2 Strategies

What is important in games is not a single move by a player, but the player *following a strategy.* A strategy for a given player is a *list of instructions* which makes a recommendation for what the player should do in every possible situation that might come up during the play of the game where it is his turn to move. To come back to our earlier example, for player A, actions (choices) and strategies come to the same thing, given that there are no other players that move before him. For the second player, B, the situation is different for he finds himself in front of tree distinct nodes. A strategy for B is a list of instructions which have a conditional form, e. g.,

- If A chooses a_1, then I will choose b_1
- If A chooses a_2, then I will choose b_2
- If A chooses a_3, then I will choose b_2 .

The strategy above gives, for each node where B is supposed to move, exactly one move. For this reason it is called a *deterministic* strategy. The games we shall consider in this book have only deterministic strategies. We shall make these informal comments more precise.

We fix an extensive game $G = (N, H, Z, P, (u_p)_{p \in N})$ with perfect information, and denote by H_p the set of histories where it is player p's turn to move, i.e. $H_p = \{h \in H - Z : P(h) = p\}$.

- A *strategy* for player p is a function σ_p which for every $h \in H_p$ yields a choice $\sigma_p(h)$ such that $h \frown \sigma_p(h) \in H$. The definition ensures that the move recommended by the strategy is a "legal" one.
- We say that player p *follows (uses) the strategy* σ_p in a history $h \in H$ if for every sub-history h' of h where it is p's turn to move (as indicated by the player function P), $h' \frown \sigma_p(h')$ is either h or an initial sub-history of h.

The second notion should be intuitively clear: player p follows the strategy σ_p in a given history, if at every node on the history where it is his turn to move, p follows the instruction prescribed by σ_p.

We note that if h has no sub-history which is a decision point for p, then it is trivially the case that p follows any strategy σ_p in h.

Example 2.1 In the example we discussed above, let σ_B be the strategy described informally, i.e. $\sigma_B(a_1) = b_1, \sigma_B(a_2) = b_2$ and $\sigma_B(a_3) = b_2$. Player B follows σ_B in the following histories:

$$\varnothing \quad (a_1), \quad (a_2), \quad (a_3), \quad (a_1, b_1), \quad (a_2, b_2), \quad (a_3, b_2)$$

where $\varnothing$ denotes the initial history (the root). Player A follows the strategy a_1 in the histories

$$\varnothing, \quad (a_1), \quad (a_1, b_1) \quad (a_1, b_2)$$

the strategy a_2 in the histories

$$\varnothing, \quad (a_2), \quad (a_2, b_1), \quad (a_2, b_2)$$

and the strategy a_3 in the histories

$$\varnothing, \quad (a_3), \quad (a_3, b_1), \quad (a_3, b_2).$$

Fix an extensive game $G = (N, H, Z, P, (u_p)_{p \in N})$ with perfect information. Suppose that each player chooses one of his strategies and follows it, starting with the root of the game. A unique play (maximal history) of the game is generated. We may take the payoffs of that history to be the payoffs that the players receive for that particular sequence of strategies (often called a *strategy profile*). Thus returning to our last example, when player A follows his strategy a_3 and player B follows the strategy σ_B described above, then the generated play is (a_3, b_2). Then:

$$u_A(a_3, sigma_B) = 2 \quad \texttt{and} \quad u_B(a_3, sigma_B) = 2.$$

We shall return to these conventions when discussing strategic games.

2.3 Solution concepts: backward induction

Example 2.2 (B. Guerrien, 2002) We consider a strictly competitive game, the New Comer game, which models the situation where a new comer (NC) to the market would like to introduce a merchandise which is already produced by an old enterprise (M, for monopoly). The options of NC are straightforward: either he comes to the market (c) or he does not come (nc). M has also two choices. If NC comes to the market, then he can "give up" (g), that is, he limits the production in order to avoid a collapse of the prices; or he does not give up (ng). There are then 3 possibles plays (maximal histories):

1. NC does not come to the market in which case M takes all the benefit $(0, 10)$
2. NC comes to the market and M gives up so that they share sales and benefits $(4, 4)$
3. NC comes to the market and M does not give up, in which case both produce but lose money $(-3, -2)$.

We can represent this game in extensive form as shown in figure 2.2.

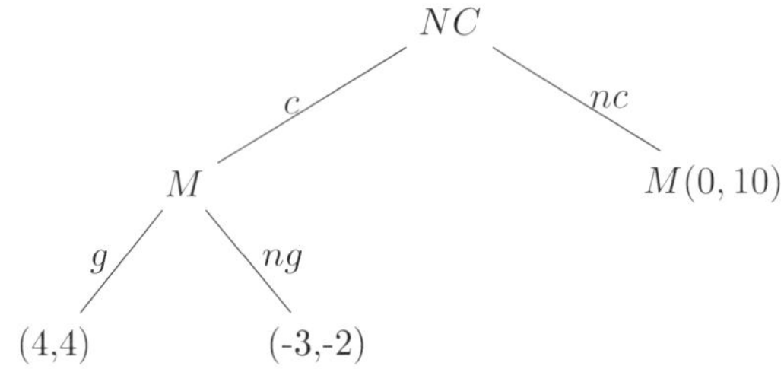

Figure 2.2: An extensive game tree for the New Comer game

We notice that NC has two strategies which coincide with his actions: c and nc. On the other side, given that it is M's turn to move only in one history of the game, that is, $H_M = \{c\}$, we can take his strategies to coincide with his actions: g and ng. How will the two players reason to determine what is rational to do? The reasoning is known as *backward induction*.

- M will reason like this: If NC decides not to enter the market, then he does not need to consider any of the two options, but he will continue his production getting the maximum benefit (10). If on the other side NC decides to enter the market, it is in his interest to give up, given that he will get a better payoff (4) than if did not give up (-2).
- NC, by putting himself in M's shoes, anticipates his reasoning, decides to enter the market, because he receives a better payoff (4) than if he did not (0).

The solution of the game is obvious: NC follows his strategy c and M follows his strategy g. When they do so, a unique play of the extensive game is generated, the second one in the list above, and each of the players receives the payoff 4.

Backward induction is the reasoning process which, as we have seen, starts from the end of the game-tree, evaluates what the player having the last move would do, then evaluates what the player having the preceding move will do, up to the root of the tree. The method leads to a reduction of the initial game-three to smaller and smaller one until the root is attended. Thus in our example, we saw that it is in M's interest to give up. This reduces the tree to the smaller one:

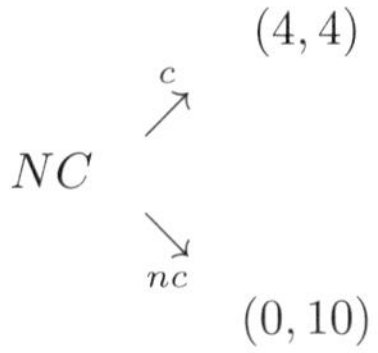

Now it is in NC's interest to choose c because he gets a better payoff. So the solution of the game is the pair of strategies (c, g). That is, the new comer decides to come to the market and the Monopoly decides to give up and accept it. The corresponding payoffs are:

$$u_{NC}(c, g) = 4 \quad \text{and} \quad u_M(c, g) = 4.$$

Example 2.3 (The Centipede, B. Guerrien, 2002) We give another example of a game where the method of Backward Induction is used. There are two players, A and B each of them having the choice between the action a (which terminates the game) and the action c (which continues it):

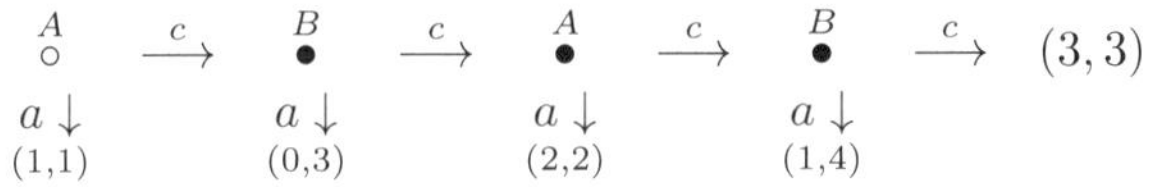

Notice that each player has the same set of four strategies: $(a, a), (a, c), (c, a)$ and (c, c). The solution by Backward Induction goes like this. At his last move, Player B being rational, will prefer a to b. At the preceding stage, player A anticipates B's reasoning at the next stage, and (assuming B is rational at the next stage), will prefer a (payoff 2) to c (payoff 1). At a still an earlier stage, player B anticipates player's A reasoning at the next stage and (assuming A is rational at that stage) will prefer a (and get payoff 3) to c (for which he will get payoff 2). Finally, at the root, by a similar reasoning A will prefer a to c. The solution of the game is: A follows the strategy (a, a) and B follows the same strategy (a, a). The play of the game generated by this pair is (a) which returns to each player payoff 1. The extensive form of the Centipede obscures the fact that there are actually four solutions to the game:

$$\{(a, a), (a, a)\}, \quad \{(a, c), (a, a)\}, \quad \{(a, a), (a, c)\}, \quad \{(a, c), (a, c)\}$$

We notice that each solution generates the same play (a) and thus returns the same payoffs to the players. We have here a good example between the *actual play* generated by a pair of strategies and the *hypothetical histories* generated by the choices of the players. A strategy for a player is a contingency plan that has to be defined also for the histories which *might* be reached. On the other side, the payoffs of the players are determined by the actual play. The Centipede is a well known play in the game-theoretical literature because the solution of the game is (Pareto) *sub-optimal*: there is another pair of strategies, $\{(c,c),(c,c)\}$ which generates the history (c,c,c,c) and returns to each player a superior payoff: 3.

2.4 Backward Induction and Common knowledge of rationality

Actually in the counterfactual reasoning we described above, there is more than meets the eyes. Let us look closer to what it involves in the Centipede Game. We run through the argument and detect the assumptions involved.

- If

1. Player B is rational in his last move,

then the game reduces to

$$\begin{array}{ccccccc} \overset{A}{\circ} & \xrightarrow{c} & \overset{B}{\bullet} & \xrightarrow{c} & \overset{A}{\bullet} & \xrightarrow{c} & (1,4) \\ a\downarrow & & a\downarrow & & a\downarrow & & \\ (1,1) & & (0,3) & & (2,2) & & \end{array}$$

- If

2. Player A knows that player B is rational at his last move,

then he can infer that the game reduces to the smaller one.

- If in addition

3. Player A is rational in his last move,

then, the game reduces to

$$\begin{array}{ccccc} \overset{A}{\circ} & \xrightarrow{c} & \overset{B}{\bullet} & \xrightarrow{c} & (2,2) \\ a\downarrow & & a\downarrow & & \\ (1,1) & & (0,3) & & \end{array}$$

- If player B knows (2) and (3), that is, if

4. Player B knows that Player A knows that player B is rational in his last move

5. Player B knows that player A is rational in his last move

then player B can infer that the game reached the smaller game.

- If in addition

6. Player B is rational in his first move

then the game is reduced to

$$\begin{array}{ccc} \overset{A}{\circ} & \xrightarrow{c} & (0,3) \\ a \downarrow & & \\ (1,1) & & \end{array}$$

- If player A knows (4), (5) and (6), that is, if

7. Player A knows that Player B knows that Player A knows that player B is rational in his last move

8. Player A knows that Player B knows that player A is rational in his last move

9. Player A knows that Player B is rational in his first move

he can infer that the smaller game has been reached.

- If in addition

10. Player A is rational at his first move

then he chooses a.

- Assumptions of the form "Player A knows that Player B knows that Player A knows that player B is rational in his last move" are captured by the assumption of Common knowledge of rationality.

The Backward Induction solution yields payoffs $(1,1)$ for the Centipede game. The motivation of this solution in terms of the assumption of Common Knowledge of rationality led to epistemic considerations and epistemic logic as one of the foundations of game theory. One of the standard results in the field is Aumann's theorem. We mention it without proof but the previous example should give an intuitive idea about the reasoning behind the proof:

Theorem 2.1 (Aumann, 1995) *The Backward Induction solution must obtain in extensive games of perfect information where players have common knowledge of rationality.*

See also von Benthem (2014, p. 23) for a discussion of this result.

2.5 Win-lose games

The following definitions are standard in the literature. We follow the exposition in Mann et al (2011).

A 2-player game is one in which N consists of two players, we denote them usually by player I and player II. A 2-player game is *strictly competitive* if the players have no incentive to cooperate, that is, for every $h, h' \in Z$:

$$u_I(h) \geq u_I(h') \quad \texttt{iff} \quad u_{II}(h') \geq u_{II}(h).$$

A 2-player game is said to be a *constant-sum* game if the players' payoff is constant, that is, there exists $c \in \mathbb{R}$ such that for each maximal history h we have

$$u_I(h) + u_{II}(h) = c.$$

When $c = 0$ (1) we say that the game is 0-sum (1-sum), etc. In a constant-sum game the interests of the players are diametrically opposed: every win for a player corresponds to a loss for the other. Obviously every constant sum game is a strictly competitive game but the converse does not hold. Counter-examples are not difficult to come by. Think of a two players extensive game with three maximal histories h_1, h_2 and h_3 such that the payoffs are

$$(1,1) \qquad (2,0) \qquad (1,2).$$

If the only possible payoffs are 0 and 1, we say that player I *wins* the maximal history h and player II *loses* h if $u_I(h) = 1$ and $u_{II}(h) = 0$. An extensive constant sum game is a *win-lose* if for each maximal history, exactly one player wins it (and thus exactly one player loses it). That is, for each $h \in Z : u_p(h) = 0$ or $u_p(h) = 1$ (but not both).

We adopt the following notational conventions:

- Z_p denotes the set of plays (maximal histories) of Z which are won by player p. (This makes sense only for win-lose games)
- H_σ is the set of histories where a strategy σ is followed.
- Z_σ is the set of plays in which a strategy σ is followed.

Obviously$Z_p \subseteq H_p$.

Let σ_p be a strategy for player p. σ_p is a *winning strategy for player p* if player p wins every play where she follows σ_p, i.e. $Z_\sigma \subseteq Z_p$.

Note once again that a strategy for a player p is defined for all possible histories where p is to move (i.e. H_σ) but in order to determine whether a given strategy is winning or not, one needs to consider only the histories where the strategy is followed. Note also that the idea behind "a player p wins every play where she follows σ_p" is that σ_p leads to a win for p against any possible move by p's opponent.

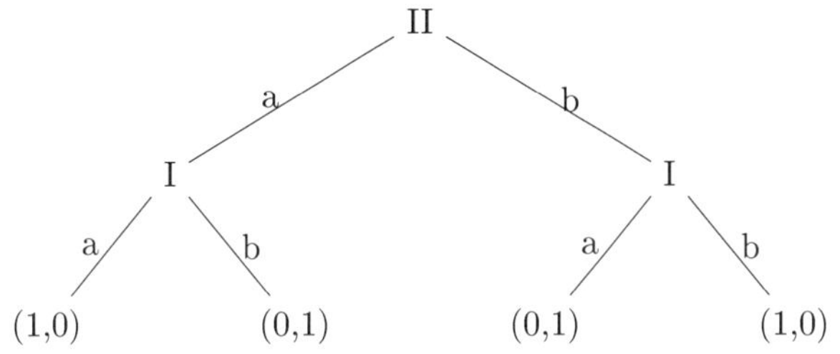

Figure 2.3: A win-lose game with perfect information

Example 2.4 Consider the extensive game with perfect information as shown in figure 2.3.

In this game player I wins the histories (a, a) and (b, b) and player II wins the histories (a, b) and (b, a). In this game player I has a winning strategy σ defined by:

$$\sigma(a) = b \quad \texttt{and} \quad \sigma(b) = a$$

Player I follows this strategy in two plays, (a, b) and (b, a) and she wins both. On the other side player II has two strategies, a and b. None of them is winning. He follows a in two plays: (a, a) and (a, b) but he loses the second. Similarly, he follows b in (b, b) and (b, a) but he also loses the second.

2.6 Determinacy of games: Zermelo's theorem

We consider a special case of the Backward Induction argument which arises for extensive, win-lose games of perfect information. It is obvious from the definitions of the notion of winning strategy in extensive, win-lose games with perfect information, that it is impossible for both players to have a winning strategy in the game. This follows simply from the fact mentioned earlier that when each player follows one of his strategies, then the result is a play of the game. But in a win-lose game, for each play exactly one of the player wins it and one of the player loses, but not both. There still remains the question whether each win-lose games of perfect information is *determined*: either one of the players has a winning strategy.

Theorem 2.2 (Zermelo) *Every two-player, win-lose, extensive game with perfect information that has finite horizon is determined.*

Proof We give an informal proof (van Benthem, 2014) Here is a simple algorithm determining the player having the winning strategy at any given node of the game tree. It works bottom-up through the game tree.

- First, colour those maximal histories with the colour Black that are wins for Abelard, and colour the other histories, the ones which are wins for

Eloise, with colour White. Given that each history is either a win for Eloise, or a win for Abelard (and no history is a win for both players), each history will receive exactly one colour.

- Extend the colouring as follows. If all children of node n have coloured already, do one of the following:

(a) If player II is to move, and at least one child is Black: colour n Black; if all children are White, colour n White.

(b) If player I is to move, and at least one child is White: colour n White; if all children are Black, colour n Black.

- Finally we reach the top (root) of the tree. It is marked either Black or White. If it is coloured White, Eloise has a winning strategy. If it is coloured Black, Abelard has a winning strategy. Here is the reasoning.

Suppose the root is coloured White. We describe a strategy σ for player I in the game. Let H_I be the set of histories where it is player I's turn to move. Suppose $h \in H_I$. Define $\sigma(h)$ to be any of the children nodes a such that $h \frown a$ is coloured White, if such a node exists. Otherwise, let $\sigma(h)$ be defined arbitrarily.

We claim that σ is a winning strategy for player I. To do so, let H_σ be the set of histories where the strategy σ is followed. We now show that every history in H_σ is coloured White and thus every maximal history in H_σ is coloured White. That is, every maximal history where player I follows σ is a winning one for her.

We already know that the root, which belongs to H_σ is coloured White. Suppose that $h \frown a \in H_\sigma$. Then $h \in H_\sigma$. By the inductive hypothesis, h is coloured White. Now if h is a decision point for player I, then given the colouring procedure, there must be a child node of h which is coloured White, and by the definition of σ, that node must be a , i.e., $\sigma(h) = a$. If on the other side h is a decision point for Abelard, then all the children nodes of h must be white, for if one of them were coloured Black, then by the colouring procedure, h should have been coloured Black too. Thus $h \frown a$ is coloured white.

The case of the top node being coloured black is completely symmetric.

As the top node is either coloured white or black, then either player I has a winning strategy or player II has a winning strategy in the game. A full proof is to be found in Mann et al (2011), Theorem 2.5. □

2.7 Extensive games with imperfect information

These games have the same structure as in the perfect information case, except for the presence of a new ingredient which codifies the information of the two players in a game. That is, a 2-player, extensive game with imperfect information is a tuple

$$G = (N, H, Z, P, \{\sim_p : p \in N\}\ (u_p)_{p \in N})$$

where N, H, Z, P and u_p are exactly as before (i.e. perfect information case). The new element is the relation $\sim_p$, *the information set* of player p, which is, technically speaking, an equivalence relation on H_p (the set of nonmaximal histories where it is player's p turn to move). The equivalence relations $\sim_p$ constitute primitive information that comes with the specification of the game. When two histories $h, h' \in H_p$ are equivalent, i.e., $h \sim_p h'$, we say that h and h' are *indistinguishable* for player p.

Graphically we indicate that two histories are indistinguishable for a player by connecting the histories with a dotted line like in the figure 2.4.

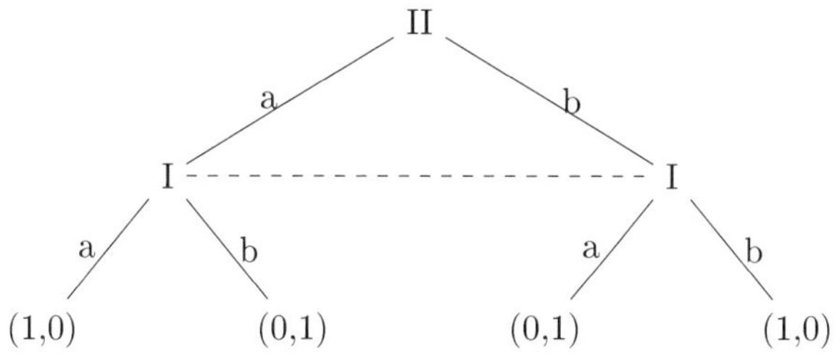

Figure 2.4: A game with imperfect information

In this example, when choosing between a or b, player II knows he is at the root of the game. The initial history is indistinguishable from itself for player II (we shall not indicate this with a dotted line.) When choosing between a and b, Player I does not know whether player II chose a or b, before him. That is, $a \sim_I b$.

Of course, for a player p to not be able to distinguish between the histories h and h', it must be that the actions available for him at either history are identical, for otherwise he would use this fact as a criterion for distinguishing the two histories. In other words, the following constraint will be imposed on the set of histories of a game:

$$h \sim_p h' \quad \Rightarrow \quad \{a : h \frown a \in H\} \quad = \quad \{a : h' \frown a \in H\}.$$

We notice that this constraint is satisfied in our example.

A strategy σ_p for a player p in an extensive game with imperfect information is a list of instructions, like in the perfect information case. There is, however one important constraint on strategies:

Uniformity A strategy σ_p for player p must make the same recommendation for histories that are indistinguishable for p:

$$h \sim_p h' \quad \Rightarrow \quad \sigma_p(h) = \sigma_p(h').$$

The notions of *competitive*, *constant-sum*, and *win-lose* game are defined exactly as in the perfect information case. Also, for a win-loose game, one can introduce the notion of winning a play of the game, and the notion of winning strategy exactly as in the perfect information case.

Example 2.5 (Matching Pennies) The game in figure 2.4 is known in the game-theoretical literature as the Matching Pennies. We may think of it as representing the following situation. Player II turns a coin to Heads or Tails without player I seeing it. After this, player I does the same. Player I wins if the outcomes match; player II wins if they differ. It is useful to compare this game to its perfect information variant described in section 2.4. Notice that, as in the earlier example, player $\forall$ (which corresponds to player II) has two strategies, a or b. But now, because of the uniformity constraint, player $\exists$ (which corresponds to player I) has two strategies instead of four:

1. If $\forall$ chooses a or b, then I will choose a :

$$\sigma_1(a) = \sigma_1(b) = a$$

2. If $\forall$ chooses a or b, then I will choose b :

$$\sigma_2(a) = \sigma_2(b) = b.$$

It is easy to see that none of the players has a winning strategy in the game. For in order for σ_1 to be a winning strategy we must have

$$a = \sigma_1(a) = \sigma_1(b) = b$$

which is impossible. The same kind of argument rules our σ_2 as a winning strategy. Symmetrically, a cannot be a winning strategy for player II, for in that case we must have:

$$a \neq a$$

which is impossible, etc. The Matching Pennies is our first example of an extensive game with imperfect information that is undetermined.

Example 2.6 Consider the extensive game of imperfect information described in the figure 2.5.

Player A has four strategies, a_1, a_2, a_3 and a_4. For player B, the histories a_1 and a_2 are equivalent, and so are the histories a_3 and a_4. Given the requirement of uniformity, player B has 4 strategies:

1. If player A chooses a_1 or a_2, then I will choose b_1 ; If player A chooses a_3 or a_4, then I will choose b_1

$$\tau_1(a_1) = \tau_1(a_2) = \tau_1(a_3) = \tau_1(a_4) = b_1.$$

2. If player A chooses a_1 or a_2, then I will choose b_1 ; If he chooses a_3 or a_4, then I will choose b_2 (σ_2) (the formal definition is similar to the above)

3. If player A chooses a_1 or a_2, then I will choose b_2 ; If he chooses a_3 or a_4, then I will choose b_1 (σ_3)

4. If player A chooses a_1 or a_2, then I will choose b_2 ; If he chooses a_3 or a_4, then I will choose b_2 (σ_4).

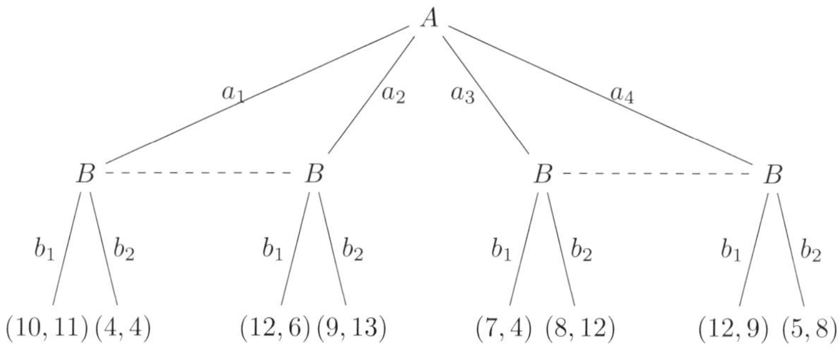

Figure 2.5: A game of imperfect information

Chapter 3

Strategic games

3.1 Basic notions

Extensive games emphasize the sequential nature of the interaction process, i.e. how a player's move depends on the moves of other players earlier in the game. A strategy for the player is, as we have seen, a list of instructions which tells him what to choose at every possible situation where it is his or her turn to move. Once the players have selected the strategy to follow in the game, the outcome of the extensive game is completely determined: it is the unique history of the game tree in which every play (maximum history) has a determined payoff.

A strategic game abstracts from the sequential structure of the extensive game and focuses, instead, on the moment when the players select the strategies to follow. Once these strategies are selected, the outcome is determined and the players receive their payoffs. For an example, let us return to the New Comer game considered earlier.

Example 3.1 To obtain the strategic form of the New Comer game, we consider the set of strategies of the two players in the extensive game that we denote by S_{NC} and S_M, respectively. In this case, as we have noticed, both sets coincide with the set of choices of the two players. When NC selects one of his strategies, say $c \in S_{NC}$, and M selects one of his strategies, say $ng \in S_M$, then a unique play of the extensive game is generated, which fixes the payoffs of the two players, -3 for NC and -2 for M. It is customary to represent a strategic game in the form of matrix which exhibits the payoffs of the players for each possible choice of strategies. Here is the New Comer game in strategic form:

	g	ng
c	$(4,4)$	$(-3,-2)$
nc	$(0,10)$	$(0,10)$

(the row player is NC and the column player is M.)

We are ready for the general definition.

Definition 3.1 A strategic game $\Gamma = (N, (S_p)_{p \in N}, (u_p)_{p \in N})$ has the following components:

- N is the set of players
- S_p is the set of all possible (pure) strategies of player p
- $u_p : \prod_{i \in N} S_i \to \mathbb{R}$ is the utility function for player p

We shall focus in this study on the case in which N consists of two players that very often we shall call player I and player II. In this case, the utility function for player p is any function

$$u_p : S_I \times S_{II} \to \mathbb{R}$$

That is, $u_I(\sigma, \tau)$ is the real number which is the payoff of player I when he plays the strategy $\sigma \in S_I$ and player II plays the strategy $\tau \in S_{II}$; and $u_{II}(\sigma, \tau)$ is the real number which is the payoff of player II.

A strategic game is *finite* if both S_I and S_{II} are finite.

The notions of strictly competitive, constant sum, zero-sum and win-lose strategic games are defined exactly as in extensive games, except that now histories are replaced by pairs of strategies. For instance, a 2 player strategic game is strictly competitive if and only if for all $\sigma, \sigma' \in S_I$ and all $\tau, \tau' \in S_{II}$ we have

$$u_I(\sigma, \tau) \geq u_I(\sigma', \tau') \quad \texttt{iff} \quad u_{II}(\sigma', \tau') \geq u_{II}(\sigma, \tau).$$

3.2 Solution concepts: The elimination of dominated strategies

The particularity of strategic game is that the players must make decisions in interaction, which means that a player must anticipate what the other players will do, before deciding which strategy to choose. There is, however, a group of games for which the optimal solution for each player is quite easy, i.e. it follows automatically from the assumption that the players are rational and that they know it. The games in this group have a common characteristic: the players have *dominant strategies*, either directly or by iteration. In the first case, the solution is obtained in a simple way without even being necessary for the players to rely on their beliefs about what the other players will do.

We first consider a simple game where player I (row) has a choice between two strategies, a_1 and a_2 and player II has a choice between 3 strategies, b_1, b_2 and b_3. Here is the matrix of the game:

	b_1	b_2	b_3
a_1	$(7,5)$	$(3,7)$	$(1,8)$
a_2	$(8,4)$	$(5,3)$	$(2,6)$

We notice that if player I chooses a_2 he will receive a payoff which is strictly better than if he chose a_1, no matter what player II does. In this case we say that a_2 is player's I *dominant* strategy, his strategy a_1 being dominated by it. We also notice that b_3 is the dominant strategy of player II. In these conditions the solution of the game is obvious: if the players are rational, then player I chooses his dominant strategy a_2 and player II chooses his dominant strategy b_3. That is, the solution of the game is the pair of strategies (a_2, b_3) which returns to the players the payoffs 2 and 6, respectively. The solution is puzzling, however, for there is another pair of strategies, (a_1, b_2) which gives the players strictly better payoffs. That is, the players could have gained more if they chose other strategies. That choice, however, is not "rational" according to the criterion established so far. The most discussed example of this "paradox" is the Prisoner's Dilemma.

Example 3.2 (Prisoner's dilemma, PD) Two suspects in a major crime are held in separate cells. There is enough evidence to convince each of them of a minor crime, unless one of them acts as an informer (finks) against the other. If they both stay quiet, each of them would be convicted of the minor offense and spend one year in prison. If only one of them finks, she will be freed and used as a witness against the other, who will spend four years in prison. If they both fink, each will spend three years in prison. Here is the game in strategic form:

	quiet	*fink*
quiet	$(-1, -1)$	$(-4, 0)$
fink	$(0, -4)$	$(-3, -3)$

Notice that PD is not a strictly competitive game given that

$$u_I(quiet, quiet) > u_I(fink, fink)$$

but

$$u_{II}(fink, fink) \not> u_{II}(quiet, quiet).$$

The Prisoner's dilemma can be solved by the dominant strategy method. It is obvious that if the first prisoner (row) chose *fink*, he would be strictly better off than if he chose *quiet*, independently of what the second prisoner would do: for if the second prisoner chose *quiet*, then he would get 0 (against -1) and if the second prisoner chose *fink*, he would get -3 (against -4). That is, the *fink* strategy is the *dominant* strategy for the first prisoner. Analogously, the *fink* strategy is the dominant for the second prisoner. In this conditions the choices of the players are obvious: *fink* for both of them, and we conclude that the solution of the game is (*fink*,*fink*).

The solution concept used in the last two examples is:

- Every player chooses his or her dominant strategy.

As the two examples show, for a given player i to assess whether one of her strategies is dominated, dominant or undominated, i does not need to know anything about her opponent's payoffs. Actually i does not even need to know that her opponents are rational. All she needs to know is her own payoffs and the fact that she is rational.

The situations modelled by PD reflect a kind of behaviour which is quite frequent in our society: the egoist behaviour in which one, by pursuing only his own interest, provokes a damaging result for the rest of us.

The solution discussed here is quite restrictive because it applies only to games which the players have a dominant strategy. We now look to a less restrictive solution concept: the *iterative elimination of dominated strategies*.

3.3 Solution concepts: The iterative elimination of dominated strategies

In the strategic game (B. Buerrien, 2002)

	b_1	b_2	b_3
a_1	$(2,4)$	$(3,7)$	$(2,5)$
a_2	$(4,6)$	$(6,5)$	$(3,7)$
a_3	$(5,4)$	$(2,3)$	$(4,2)$

there is no dominant strategy for either player. But we notice that the strategy a_1 is (strictly) dominated by the strategy a_2 for player I.

Thus, if player I is rational he will not choose a_1. We can thus eliminate the line a_1 from the matrix and reduce the game to

	b_1	b_2	b_3
a_2	$(4,6)$	$(6,5)$	$(3,7)$
a_3	$(5,4)$	$(2,3)$	$(4,2)$

Player II knows that player I is rational. He (as well as player I) can infer that the solution of the game will be found from smaller one. In this smaller game, the strategy b_2 is dominated by b_1 for player II. As Player II is rational he won't choose b_1 and both players (knowing each of them is rational) infer that the solution of the game will be found from the smaller game:

	b_1	b_3
a_2	$(4,6)$	$(3,7)$
a_3	$(5,4)$	$(4,2)$

In this smaller game, the strategy a_3 is dominant for player I. Eliminating it reduces the game to

	b_1	b_3
a_3	$(5,4)$	$(4,2)$

where player II choose b_1. The solution of the game is thus (a_3, b_1). The *iterative elimination of strictly dominated strategies* led to a unique solution. This method is the counterpart in strategic games of the backwards induction method in extensive games. We have already pointed out that the assumption that each player knows that the other player as well as himself is rational, and each player knows that the other player knows that he is rational, and so on, ad infinitum, is known as the assumption of common knowledge of rationality.

In some examples it is worth considering a weaker notion of domination. If σ and σ' are strategies for a player p, then:

- σ *dominates* σ' if for every strategy τ of the other player, we have $u_p(\sigma,\tau) > u(\sigma',\tau)$.
- σ *weakly dominates* σ' if for every strategy τ of the other player, we have $u_p(\sigma,\tau) \geq u(\sigma',\tau)$ and there is at least one strategy σ'' of player p such that $u_p(\sigma,\tau) > u(\sigma'',\tau)$.

We review a couple of examples where the weaker notion is needed.

Example 3.3 Recall the strategic form of the New Comer game:

	g	ng
c	$(4,4)$	$(-3,-2)$
nc	$(0,10)$	$(0,10)$

We observe that the strategy g weakly dominates the strategy ng for player M. Eliminating the weakly dominated strategy reduces the game to

	g
c	$(4,4)$
nc	$(0,10)$

In this smaller game, the strategy c dominates the strategy nc. The solution of the game is thus (c,g), the same one we found by backward induction.

Example 3.4 (Guerrien, 2002) Here is the strategic form of the Centipede game:

	(a,a)	(a,c)	(c,a)	(c,c)
(a,a)	$(1,1)$	$(1,1)$	$(1,1)$	$(1,1)$
(a,c)	$(1,1)$	$(1,1)$	$(1,1)$	$(1,1)$
(c,a)	$(0,3)$	$(0,3)$	$(2,2)$	$(2,2)$
(c,c)	$(0,3)$	$(0,3)$	$(1,4)$	$(3,3)$

(A is the row player). The strategy (c,a) weakly dominates the strategy (c,c). for player B. Eliminating the weakly dominated strategy reduces the game to the smaller one:

	(a,a)	(a,c)	(c,a)
(a,a)	$(1,1)$	$(1,1)$	$(1,1)$
(a,c)	$(1,1)$	$(1,1)$	$(1,1)$
(c,a)	$(0,3)$	$(0,3)$	$(2,2)$
(c,c)	$(0,3)$	$(0,3)$	$(1,4)$

In this smaller game A's strategy (c, c) is weakly dominated by each of the other three strategies. This reduces the game to

	(a,a)	(a,c)	(c,a)
(a,a)	$(1,1)$	$(1,1)$	$(1,1)$
(a,c)	$(1,1)$	$(1,1)$	$(1,1)$
(c,a)	$(0,3)$	$(0,3)$	$(2,2)$

where$B$'s strategy $(c.a)$ is weakly dominated by the other two. This further reduces the game to

	(a,a)	(a,c)
(a,a)	$(1,1)$	$(1,1)$
(a,c)	$(1,1)$	$(1,1)$
(c,a)	$(0,3)$	$(0,3)$

where$A$'s strategy (c, a) is dominated by the other two. Finally we end up with the game

	(a,a)	(a,c)
(a,a)	$(1,1)$	$(1,1)$
(a,c)	$(1,1)$	$(1,1)$

We recover the 4 solutions to the game that we found by backward induction. Notice that they all stop the game after the first move by A. The solution by iterative elimination of (weakly) dominated strategies can be seen to be better then the backward induction method, for it allows us to see that there are 4 solutions to the game (even if they give the same payoffs). Here, as in the Prisoner's dilema game, the solution is suboptimal, because there are other solutions which give better payoffs to the players.

Recall Aumann' theorem and the relation between the assumption of common knowledge of rationality and Backward Induction in extensive games. A similar relation obtains between the same assumption and the iterated elimination of (weakly) dominated strategies in strategic games. That is, it can be shown, how, by assuming that the rationality of the players is common knowledge, we can use the method of the iterated elimination of dominated strategies to reduce the game to a smaller one in which the set of strategies of each player has survived the elimination procedure. The conclusion is that under that assumption, no player would choose a strategy outside her surviving set. We will not do this here but refer the reader to Jim Rattliff (1997).

3.4 Solution concepts: Equilibria and best response strategies

The games we considered until now could be solved by backward induction (iterative elimination of dominated and weakly dominated strategies). But the games we consider in this section cannot be solved this way (neither is there one dominating strategy). In this case we are looking for a more restrictive solution concept. We start by examining for each strategy pair (profile) (a_i, b_j), each player's *best response* from the point of view of his rationality. In the strategic game (Guerrien, 2002)

	b_1	b_2	b_3
a_1	$(2,2)$	$(1,6)$	$(4,4)$
a_2	$(3,3)$	$(2,2)$	$(2,0)$
a_3	$(2,1)$	$(7,5)$	$(2,7)$

we look at the strategy profile (a_1, b_3). If player I (row) plays a_1 and player II plays b_3, then one of the players, II, will regret *ex post* his choice: if he he knew that player I would choose a_1 he would have chosen b_2! It may be checked that the players regret their choices for all the other strategy profiles, except one: (a_2, b_1). When the game is played in this way, none of the players regrets the choice he made. We say that the strategy profile (a_2, b_1) is an equilibrium in the game.

We give the formal definition of an equilibrium for 2-player strategic games $\Gamma = (N, (S_p)_{p\in N}, (u_p)_{p\in N})$ where $N = \{I, II\}$.

Definition 3.2 Let $\Gamma = (\{I, II\}, S_I, S_{II}, u_I, u_{II})$ be a strategic 2 player game such that $\sigma \in S_I$ and $\tau \in S_{II}$. The pair (σ, τ) is an *equilibrium* if the following (no regret) conditions hold:

1. For every strategy $\sigma' \in S_I$, $u_I(\sigma, \tau) \geq u_I(\sigma', \tau)$, that is,

$$u_I(\sigma, \tau) = max_{\sigma'\in S_I} u_I(\sigma', \tau)$$

2. For every strategy $\tau' \in S_{II}$, $u_{II}(\sigma, \tau) \geq u_{II}(\sigma, \tau')$, that is,

$$u_{II}(\sigma, \tau) = max_{\tau'\in S_{II}} u_{II}(\sigma, \tau').$$

The "no regret" condition (1) can be described alternatively by saying that σ is a *best response* of player I to the strategy τ of player II. Similarly condition (ii) says that the strategy τ is a best response of player II to the the strategy σ of player I. We notice that when a player plays a best response to the strategy choice of the other player, there is no *unilateral deviation* for him that will bring him a strictly higher payoff.

There is a visual technique we can use to identify strategy profile equilibria in a finite 2-player strategic game, when they exist (cf. Mann et al 2011):

- In each column circle the maximum payoffs for player I (if there are several of them, circle them all).
- In each row circle the maximum payoff for player II.

In our last example, we circle 3 in the first column, 7 in the second, and 4 in the third. We circle 6 in the first row, 3 in the second, and 7 in the third. A pair of strategies (σ, τ) is an equilibrium if and only if both $u_I(\sigma, \tau)$ and $u_{II}(\sigma, \tau)$ are circled. In our example this holds only for the pair (a_2, b_1). This visual technique is nothing else but a variant of the argument saying that each player is playing a best response to the strategy choice of the other.

Recall that if Γ is a strictly competitive game, then for all $\sigma, \sigma' \in S_I$ and all $\tau, \tau' \in S_{II}$ we have

$$u_I(\sigma, \tau) \geq u_I(\sigma', \tau') \quad \texttt{iff} \quad u_{II}(\sigma', \tau') \geq u_{II}(\sigma, \tau).$$

But then clause (2) in the definition of an equilibrium may be rewritten as:

2*. For every strategy $\tau' \in S_{II}$, $u_I(\sigma, \tau') \geq u_I(\sigma, \tau)$

$$u_i(\sigma, \tau) = min_{\tau' \in S_{II}} u_I(\sigma, \tau').$$

Thus for a strictly competitive game the strategy profile (σ, τ) is an equilibrium if and only if for every $\sigma' \in S_I$ and every $\tau' \in S_{II}$ we have:

$$max_{\sigma' \in S_I} u_I(\sigma', \tau) = u_I(\sigma, \tau) = min_{\tau' \in S_{II}} u_I(\sigma, \tau').$$

There are 2-player strategic games that do not have an equilibrium. A well known example is the *Matching Pennies* game discussed earlier

	heads	*tails*
heads	(1, 0)	(0, 1)
tails	(0, 1)	(1, 0)

which is a strictly competitive (actually a win-lose) game.

On the other side, in so-called *coordination* and *anti-coordination* games there are at least two equilibria. A standard example of a coordination game is the *Battle of the sexes*. A man and a woman may meet at a football match or at a ballet performance. Given that they want to be together, they get a higher payoff if they meet than if they don't. The woman, however, prefers to meet at the ballet whereas the man prefers to meet at the football match:

	Ballet	*Football*
Ballet	(2, 1)	(0, 0)
Football	(0, 0)	(1, 2)

There are two equilibria. The equilibrium $(Ballet, Ballet)$ gives the row player (the woman) a slightly better payoff than the other equilibrium $(Football, Football)$ which favours the column player (the man). We notice that this is not a strictly competitive game given that

$$u_I(Ballet, Ballet) > u_I(Football, Ballet)$$

but

$$u_{II}(Football, Ballet) \not\geq u_{II}(Ballet, Ballet).$$

The labelling *coordination game* has to do with the fact that the players, even if they have diverging interests, gain more if they agree than if they do not. We shall see more examples of coordination games (The Stag Hunt game) later on.

The *anti-coordination game of the chicken* or the *hawk-dove* game

	swerve	*straight*
swerve	$(0,0)$	$(-1,1)$
straight	$(1,-1)$	$(-10,-10)$

has also two equilibria where one player swerves while the other does not: $(straight, swerve)$ and $(swerve, straight)$. This game models situations of conflict in game theory in which there is competition for shared resources. The typical example is the situation of two drivers who drive towards each other: one must swerve, or both may die in the crash, but if one driver swerves and the other does not, the one who swerved will be called a "chicken," (meaning a coward). Notice that this is not a strictly competitive game, given that

$$u_I(straight, swerve) > u_I(swerve, swerve)$$

but

$$u_{II}(swerve, swerve) \not> u_{II}(straight, swerve).$$

Notice that, unlike in the coordination game discussed earlier, where playing different strategies makes both players worse, in the present case, playing the same strategy (straight) makes both players worse.

The notion of equilibrium applies to extensive games as well. We just convert an extensive game to a strategic game.

3.5 Equilibria and players' beliefs: rationalization

When discussing the iterated elimination of strictly dominated strategies, the conclusion was that common knowledge of the rationality of the players implies that the game outcome's must survive the iterated elimination process. But this does not imply that every strategy that is in the survival set could be chosen by a rational player. To illustrate what we have in mind, consider the strategic game (I am indebted to Ratliff, (1977) that I will follow closely; a nice discussion may also be found in Granoff & Lehtinen, (2014)):

	w	x	y	z
A	$(7,5)$	$(-8,4)$	$(0,4)$	$(99,3)$
B	$(5,0)$	$(4,1)$	$(15,9)$	$(100,8)$
C	$(6,0)$	$(5,8)$	$(20,4)$	$(10,2)$
D	$(2,6)$	$(7,-10)$	$(3,9)$	$(10,8)$
E	$(1,6)$	$(2,-10)$	$(1,7)$	$(8,6)$

We notice that strategy z is strictly dominated by y and thus, assuming common knowledge of rationality (cf. our earlier discussion), z may be eliminated:

	w	x	y
A	$(7,5)$	$(-8,4)$	$(0,4)$
B	$(5,0)$	$(4,1)$	$(15,9)$
C	$(6,0)$	$(5,8)$	$(20,4)$
D	$(2,6)$	$(7,-10)$	$(3,9)$
E	$(1,6)$	$(2,-10)$	$(1,7)$

In this smaller game, B is dominated by C which further reduces the game to

	w	x	y
A	$(7,5)$	$(-8,4)$	$(0,4)$
C	$(6,0)$	$(5,8)$	$(20,4)$
D	$(2,6)$	$(7,-10)$	$(3,9)$
E	$(1,6)$	$(2,-10)$	$(1,7)$

Now E is dominated by both C and D, which further reduces the game to:

	w	x	y
A	$(7,5)$	$(-8,4)$	$(0,4)$
C	$(6,0)$	$(5,8)$	$(20,4)$
D	$(2,6)$	$(7,-10)$	$(3,9)$

We can check that there is no dominated strategy which could be eliminated any longer. As already pointed out, this does not mean that it would be rational for the players to choose *any* of the remaining pairs. Given that the strategies in the smallest game have survived the iterated elimination of dominated strategies under the assumption of common knowledge, they are all consistent with this assumption. Actually each of them can be rationalized, that is, presented as a consistent system of beliefs which justifies their choices. However, only some of the pairs (actually exactly one in this case) are such that it is "rational" for the players to choose them (best response for both players). Let us show this.

Consider first the strategy profile (C, x) in the smallest game above.

A rational player I would choose C if he believed that player II would choose y. And a rational player II would play y if he believed that player I would choose D. This would happen if player I believed player II will play x. Player II would play x if he believed player I will play C. Here is the justification chain or *rationalization* for player I playing C:

1. Player I plays C
2. Player I believes that player II will play y
3. Player I believes that player II believes that player I will play D
4. Player I believes that player II believes that player I believes that player II will play x

5. Player I believes that player II believes that player I believes that player II believes that player I will play C

Notice that the rationalization process for player I playing C shows what kind of beliefs a player has to hold in order to make his choice. A similar justificatory chain can be built for player II playing x. Player II would play x if he believed that player I will play C. This would happen if player I believed player II will play y. But player II would play y if he believed that player will play D. This would happen if player I believed that player II will play x. We reached the end of the justificatory cycle. More systematically:

6. Player II plays x

7. Player II believes that player I will play C

8. Player II believes that player I believes that player II will play y

9. Player II believes that player I believes that player II believes that player I will play D

10. Player II believes that player I believes that player II believes that player I believes that player II will play x.

Now suppose that after each player rationalizes his own beliefs in the way described above and announces his own decision, that is, player I announces C and player II announces x, the players "confront" their justifications. Player II will see that he has correctly predicted both player I's choice and beliefs. Indeed, by consulting (7) we see that he correctly predicted that player I will play C; and (8-10) show that player II has correctly identified also player I's higher-order beliefs.

Things are different for player I. (2) shows that he believed incorrectly that player II will play y when actually player II played x. He also misidentified player II's higher-order beliefs. For instance, (3) shows that he believed that player II believes that player I will play D. But as shown by (7), player II believes that player I will play C.

At the end of day, when the players announce their choices, player II, who correctly justified his choice and predicted correctly player I's choice, C, does not regret his choice x : it is a best response to C. On the other side, player I regrets his choice C because he realizes he would have been better off by playing C. This comes as no surprise as he misidentified player II's choice and beliefs.

Things are different with the profile equilibrium (A, w). Here is rationalization process for player I:

11. Player I plays A

12. Player I believes that player II will play w

13. Player I believes that player II believes that player I will play A

The rationalisation process for player II is:

14. Player II plays w

15. Player II believes that player I will play A

16. Player II believes that player I believes that player II will play w

Now when the players confront their justifications, we can see that each of them correctly predicted both the other player's choice and beliefs. Given that each of them conjectured correctly the other player's beliefs, and given the fact that he plays a best strategy response based on his beliefs about what the other player will do, none of the players regrets his choice: A is player I's best response to w and w is player II's best response to A. The fact that (A, w) is, unlike (D, x), encodes a best strategy response for each player exemplifies once more our algorithm for finding equilibria in finite 2 player strategic games (when they exist): 7 is the encircled payoff for player I in the first column, and w is the encircled payoff in the first row. What the algorithm does not show is the rationalization process and the signifiance of the correct anticipation of the beliefs about what the other players might do. In our example, given that the players beliefs are correct, we can summarize the systems of beliefs (11)-(16) by saying that it is common knowledge that player I will play A and player II will play w.

Chapter 4

Mixed strategies

4.1 Mixed strategies and expected utilities

Not every game has an equilibrium in pure strategies. We have already discussed an example, the Matching Pennies game:

	heads	*tails*
heads	(1,0)	(0,1)
tails	(0,1)	(1,0)

The *Inverted Matching Pennies* game

	heads	*tails*
heads	(0,1)	(1,0)
tails	(1,0)	(0,1)

in which the row player wins if the choices of the two players do not match, and the column player wins if they match, is also a case in point. No matter which strategy pair is considered, one of the players is able to improve his payoff by an unilateral deviation. Take for instance the pair (*heads*, *tails*) in the first game. When the players reveal simultaneously their choices, it is obvious that player I will regret his choice: she could have been better off is she chose *tails*. In order to obtain a solution to the game, the players randomize their strategies.

Let Γ be a finite 2-player strategic game

$$\Gamma = (\{I, II\}, S_I, S_{II}, u_I, u_{II}).$$

A *probability distribution* over the set S_p, $p \in \{I, II\}$, is a function $\mu : S_p \to [0,1]$, that is, a function which takes as arguments the pure strategies in the set S_p and as values real numbers in the open interval $[0,1]$. In addition it is required that the values of the function sum up to 1:

$$\sum_{\sigma \in S_p} \mu(\sigma) = 1.$$

We denote by $\Delta(S_p)$ the set of all probability distributions over the set S_p.

A *mixed strategy* for player p is a probability distribution $\mu \in \Delta(S_p)$. In this book we shall consider only probability distributions over sets S_p that are finite. For a mixed probability distribution μ over a finite S_p, we call the *support* of μ the set of pure strategies in the set S_p to which μ assigns non-zero probability. We say that μ is *uniform* of it assigns equal probability to all the pure strategies in its support.

Let Γ be a strategic 2 player game, and let $\mu \in \Delta(S_I)$ and $\nu \in \Delta(S_{II})$ be mixed strategies for the two players. For σ a pure strategy for player I, i.e. $\sigma \in S_I$, and τ a pure strategy for player II, i.e., $\tau \in S_{II}$, the probability that player I follows the strategy σ is $\mu(\sigma)$ and the probability that player II follows the strategy τ is $\nu(\tau)$. We now compute the expected utility for player p for a mixed strategy profile (μ, ν).

Let $\sigma \in S_I$ be a pure strategy for player I and ν a mixed strategy for player II.

Player I's expected utility for playing σ against player II's mixed strategy ν, denoted by $U_I(\sigma, \nu)$, is given by:

$$U_I(\sigma, \nu) = \sum_{\tau \in S_{II}} \nu(\tau) u_I(\sigma, \tau)$$

Analogoulsy, when $\tau \in S_{II}$ is a pure strategy for player II and μ a mixed strategy for player I, Player II's expected utility for playing τ against player I's mixed strategy μ, denoted by $U_{II}(\mu, \tau$, is given by:

$$U_{II}(\mu, \tau) = \sum_{\sigma \in S_I} \mu(\sigma) u_{II}(\sigma, \tau)$$

We are now in the position to compute the players's expected utilities for the mixed strategy profile (μ, ν).

Player I's expected utility for choosing the mixed strategy $\mu \in \Delta(S_I)$ given that player II chooses the mixed strategy $\nu \in \Delta(S_{II})$ is the weighted sum of his expected utilites $U_{II}(\sigma, \nu)$ for each pure strategy $\sigma \in S_{II}$, that is:

$$U_I(\mu, \nu) = \sum_{\sigma \in S_I} \mu(\sigma) U_I(\sigma, \nu) = \sum_{\sigma \in S_I} \sum_{\tau \in S_{II}} \mu(\sigma) \nu(\tau) u_I(\sigma, \tau)$$

Analogously, Player II's expected utility for choosing the mixed strategy ν given that player I chooses the mixed strategy μ is the weighted sum of his expected utilities $U_{II}(\mu, \tau)$ when $\tau \in S_{II}$:

$$U_{II}(\mu, \nu) = \sum_{\tau \in S_{II}} \nu(\tau) U_{II}(\mu, \tau) = \sum_{\sigma \in S_I} \sum_{\tau \in S_{II}} \mu(\sigma) \nu(\tau) u_2(\sigma, \tau).$$

Example 4.1 Recall the Matching Pennies game in strategic form:

	heads	*tails*
heads	$(1, 0)$	$(0, 1)$
tails	$(0, 1)$	$(1, 0)$

(we represent wins by 1 and losses by 0). A mixed strategy for player I is a probability distribution μ over the set of pure strategies $S_I = \{heads, tails\}$ such that if $\mu(heads) = p$, then $\mu(tails) = 1 - p$. Analogously, a probability distribution for player II is a probability distribution ν over the same set $S_{II} = S_I$ such that if $\nu(heads) = q$ then $\nu(tails) = 1 - q$. Let μ be the mixed strategy for player I such that $p = 2/3$ and ν the mixed strategy for player II such that $q = 1/4$.

Player I's expected utilities for the two pure strategies *heads* and *tails* when player II plays the mixed strategy ν, are given by (we use obvious abbreviations):

$$U_I(h, \nu) = 1/4 \cdot 1 + 3/4 \cdot 0 = 1/4$$

and

$$U_I(t, \nu) = 1/4 \cdot 0 + 3/4 \cdot 1 = 3/4$$

respectively. Hence player I's expected utility for the mixed strategy profile (μ, ν) is:

$$U_I(\mu, \nu) = \mu(h)U_I(h, \nu) + \mu(t)U_I(t, \nu) = 2/3 \cdot 1/4 + 1/3 \cdot 3/4 = 5/12.$$

Player II's expected utilities for the two pure strategies *heads* and *tails* when player I plays the mixed strategy μ, are given by:

$$U_{II}(\mu, h) = 2/3 \cdot 0 + 1/3 \cdot 1 = 1/3$$

and

$$U_{II}(\mu, t) = 2/3 \cdot 1 + 1/3 \cdot 0 = 2/3$$

respectively. Hence player II's expected utility for the mixed strategy profile (μ, ν) is:

$$U_{II}(\mu, \nu) = \nu(h)U_{II}(\mu, h) + \nu(t)U_{II}(\mu, t) = 1/4 \cdot 1/3 + 3/4 \cdot 2/3 = 7/12.$$

In the general case in which μ is the mixed strategy for player I, $\mu(heads) = p$, and ν is the mixed strategy for player II, $\nu(heads) = q$, we have:

$$\begin{aligned} U_I(h, \nu) &= q \cdot 1 + (1 - q) \cdot 0 = q \\ U_I(t, \nu) &= q \cdot 0 + (1 - q) \cdot 1 = 1 - q \end{aligned}$$

The expected utility for player I is then:

$$U_I(\mu, \nu) \quad = \quad \mu(h)U_I(H, \nu) + \mu(t)U_I(T, \nu) = pq + (1 - p)(1 - q) = 2pq - p - q + 1.$$

Symmetrically, it may be checked that

$$U_{II}(\mu, \nu) = \quad p + q - 2pq.$$

Let us finally take note of the fact that if the strategic game Γ is a finite c-sum game, that is, if

$$u_I(\sigma, \tau) + u_{II}(\sigma, \tau) \quad = \quad c$$

for every $\sigma \in S_I$ and every $\tau \in S_{II}$, then it is straightforward to show that

$$U_I(\sigma, \tau) + U_{II}(\sigma, \tau) \quad = \quad c.$$

There is another perspective on mixed strategies that is worth mentioning. We restrict ourselves to a finite 2-player strategic game. We may think of a mixed strategy ν for player II over S_{II} as a (probabilistic) belief of player I. That is, for $tau \in S_{II}$, we may identify $\nu(tau)$ with the probability that player I assigns to player II playing tau. Analogously, we may identify a mixed strategy μ for player I over S_I with a belief of player II.

In the Matching Pennies example, consider the following mixed strategy $\nu(h) = 1/3$ and $\nu(t) = 2/3$ for player II (Column). Thus if player 1 (Row) play *heads* he expects a payoff 1 with probability $1/3$ and a payoff 0 with probability $2/3$. Such a sequence is called in the literature an *expected lottery* for player I's playing *heads*. Thus player I's different actions lead to different expected lotteries.

4.2 Mixed strategy equilibria

The definition of a mixed strategy equilibrium is completely analogue to the pure strategy equilibrium.

Definition 4.1 Let Γ be a finite 2-player strategic game $\Gamma = (\{I, II\}, S_I, S_{II}, u_I, u_{II})$. The pair (μ, ν), where μ is a mixed strategy for player I and ν is a mixed strategy for player II is an equilibrium, if the following two conditions hold:

(i) for every mixed strategy $\mu' \in \Delta(S_I)$, $U_I(\mu, \nu) \geq U_I(\mu', \nu)$,that is

$$U_I(\mu, \nu) = max_{\mu' \in \Delta(S_I)} U_I(\mu', \nu).$$

(ii) for every mixed strategy $\nu' \in \Delta(S_{II})$, $U_{II}(\mu, \nu') \geq U_{II}(\mu, \nu)$, that is

$$U_{II}(\mu, \nu) = max_{\nu' \in \Delta(S_{II})} U_{II}(\mu, \nu').$$

Note that if Γ is strictly competitive, then (μ, ν) is an equilibrium if and only if:

$$max_{\mu' \in \Delta(S_I)} U_I(\mu', \nu) = U_I(\mu, \nu) = min_{\nu' \in \Delta(S_{II})} U_I(\mu, \nu').$$

Example 4.2 We continue the analysis of the Matching Pennies game from the last example and observe that the pair of mixed strategies (μ, ν) according to which player I plays *heads* with probability $2/3$ and player II plays *heads* with probability $1/4$ is not an equilibrium. For instance, player I can be better off by unilateral deviation: $\mu'(heads) = 1/2 = \mu'(tails)$. In this case:

$$\begin{aligned} U_I(\mu', \nu) &= 2pq - p - q + 1 \\ &= 2 \cdot 1/2 \cdot 1/4 - 1/2 - 1/4 + 1 \\ &= 1/2 \geq 5/12 \end{aligned}$$

Actually the pair of uniform mixed strategies (μ', ν') where μ' is like above and $\nu' = \mu'$ is an equilibrium. To see this, we notice first that for any ν :

$$\begin{aligned} U_I(\mu', \nu) &= 2pq - p - q + 1 \\ &= q - 1/2 - q + 1 \\ &= 1/2 \\ &= U_{II}(\mu', \nu') \end{aligned}$$

And similarly, for any μ:

$$\begin{aligned} U_{II}(\mu, \nu') &= 1 - U_I(\mu, \nu') \\ &= p + q - 2pq \\ &= 1/2 \\ &= U_I(\mu, \nu'). \end{aligned}$$

Finally to show that (μ', ν') is an equilibrium, we check that clauses (i) and (ii) in the definition of an equilibrium hold. By the last equation, for any mixed strategy μ of player I, we have $U_I(\mu, \nu') = 1/2$. This establishes condition (i) of the definition. By the previous equation, for any mixed strategy ν of player II, $U_{II}(\mu', \nu) = 1/2$. This establishes condition (ii). We also notice that for all strategies μ of player I and strategies ν of player II, the pairs (μ', ν) and (μ, ν') are also mixed strategy equilibria. All these equilibria return the same expected utilities to the two players, which is $1/2$.

Example 4.3 (Guerrien, 2002) Two enterprises have to decide whether they will launch the production of a new merchandise. There are two choices: *pr* (produce) and *npr* (does not produce). The payoffs of the game are described in the matrix below:

	pr	*npr*
pr	$(-3, -2)$	$(10, 0)$
npr	$(0, 8)$	$(0, 0)$

As in the previous example, we let μ be a mixed strategy for enterprise I, $\mu(pr) = p$ (and thus $\mu(npr) = 1 - p$.); and we let $\nu(pr) = q$ (and thus $\nu(npr) = 1 - q$.) Then:

$$\begin{aligned} U_I(\mu, \nu) &= pq(-3) + p(1-q)10 + (1-p)q0 + (1-p)(1-q)0 \\ &= -3pq + 10 \cdot p - 10pq \\ &= p(10 - 13q). \end{aligned}$$

Similarly

$$\begin{aligned} U_{II}(\mu, \nu) &= pq(-2) + p(1-q)0 + (1-p)q8 + (1-p)(1-q)0 \\ &= -2pq + 8q - 8pq \\ &= q(8 - 10p). \end{aligned}$$

We notice that (npr, pr) is an equilibrium in pure strategies, or, equivalently, the pair (μ^*, ν^*) such that $\mu^*(npr) = 1$, $\mu^*(pr) = 0$, and $\nu^*(pr) = 1$, $\nu^*(npr) = 0$ is a mixed strategy equilibrium. This should be obvious, but we run the argument nevertheless. Given the values of μ^*, $U_I(\mu^*, \nu^*) = 0$, as expected. And given the values of ν^*, $U_{II}(\mu^*, \nu^*) = 8$. Thus we recovered the payoffs of the pure strategy profile (npr, pr).

To establish that (μ^*, ν^*) is an equilibrium, we must still show that conditions (i) and (ii) are fulfilled. Let μ be any mixed strategy of the row player. Then $U_I(\mu, \nu^*) = -3p$. But $0 \leq p \leq 1$ from which it follows that $U_I(\mu, \nu^*) \leq U_I(\mu^*, \nu^*)$. Condition (ii) is established in a similar way. A similar argument shows that (pr, npr) is also a pure strategy equilibrium. Finally, let us show that switching to mixed strategies introduces a new, proper, mixed strategy equilibrium: $q = 10/13$ and $p = 8/10 = 4/5$. So let (μ^+, q^+) be the pair of strategies such that $\mu^+(pr) = 8/10$ and $\nu^+(pr) = 10/13$. Then $U_I(\mu^+, \nu^+) = 0$ and $U_{II}(\mu^+, \nu^+) = 0$. Actually $U_I(\mu, \nu^+) = 0$ for any μ and this establishes Condition (i). And $U_I(\mu^+, \nu) = 0$ for any ν, which establishes Condition (ii).

Wo take stock.

- We have encountered games like the Battle of the Sexes which have multiple pure strategy equilibria (and multiple mixed strategy equilibria). These are coordination games.
- We encountered games like the Matching Pennies, which have no pure strategy equilibria but plenty of mixed strategy equilibria. However, in this case, all the equilibria pairs returned the same expected utilities to the players.
- Finally, we encountered games like in our last example, which have a pure strategy equilibrium and a mixed strategy equilibrium which return distinct expected utilities to the players

The case of the Matching Pennies game illustrates a more general phenomenon.

Proposition 4.4 Let (μ, ν) and (μ', ν') be two mixed strategy equilibria in a strictly competitive game. Then

$$U_p(\mu, \nu) \quad = \quad U_p(\mu', \nu')$$

PROOF Proof: Recall that in a strictly competitive game (μ^*, ν^*) is an equilibrium in mixed strategies if and only if

$$U_I(\mu, \nu^*) \quad \leq \quad U_I(\mu^*, \nu^*) \quad \leq \quad U_I(\mu^*, \nu)$$

for every mixed strategies $\mu \in \Delta(S_I)$ and $\nu \in \Delta(S_{II})$. Given that (μ', ν') is an equilibrium in the given strictly competitive game, then we have:

$$U_I(\mu, \nu') \quad \leq \quad U_I(\mu', \nu') \quad \leq \quad U_I(\mu', \nu)$$

for μ and ν the strategies in the equilibrium pair (μ, ν). And given that (μ, ν) is an equilibrium, we also have: $U_I(\mu', \nu) \leq U_I(\mu, \nu)$ and $U_I(\mu, \nu) \leq U_I(\mu, \nu')$. Putting all this together we get:

$$U_I(\mu, \nu) \leq U_I(\mu, \nu') \quad \leq \quad U_I(\mu', \nu') \quad \leq \quad U_I(\mu', \nu) \quad \leq U_I(\mu, \nu)$$

hence $U_I(\mu, \nu) = U_I(\mu', \nu')$. A similar argument shows $U_{II}(\mu, \nu) = U_{II}(\mu', \nu')$.□

The previous proposition tells us that any two mixed strategy equilibria give the same payoffs to the two players, but does not guarantee that an equilibrium exists. Von Neumann proved that for a special class of games, there always exists an equilibrium.

Theorem 4.2 (von Neumann's Minimax Theorem) *Every finite, two person, zero-sum game has an equilibrium in mixed strategies.*

4.3 Identifying mixed strategy equilibria

We review one basic fact about 2-player, finite, strategic games which will help us to identify equilibria, that is, it allows us, given a pair (μ^*, ν^*) of strategies, to determine where this pair is an equilibrium or not. The next proposition shows it is sufficient to compare (μ^*, ν^*) against any pairs of the form (σ, ν^*) and (μ^*, τ) where σ and τ are pure strategies.

Theorem 4.3 (Osborne 2003) *Let μ^* be a mixed strategy for player I and ν^* a mixed strategy for player II in a finite, two player strategic game Γ. The pair (μ^*, ν^*) is an equilibrium in Γ if and only if the following conditions hold:*

1. $U_I(\mu^*, \nu^*) = U_I(\sigma, \nu^*)$ *for every* $\sigma \in S_I$ *in the support of* μ^*
2. $U_{II}(\mu^*, \nu^*) = U_{II}(\mu^*, \tau)$ *for every* $\tau \in S_{II}$ *in the support of* ν^*
3. $U_I(\mu^*, \nu^*) \geq U_I(\sigma, \nu^*)$ *for every* $\sigma \in S_I$ *outside the support of* μ^*
4. $U_{II}(\mu^*, \nu^*) \geq U_{II}(\mu^*, \tau)$ *for every* $\tau \in S_{II}$ *outside the support of* ν^*

In other words, in other to establish whether the pair (μ^*, ν^*) is an equilibrium, it is enough to compare it only against all the pairs (σ, ν^*), where σ is a pure strategy for player I. The cases (2) and (4) are symmetrical.

PROOF The proof may be found in Osborne 2003, p. 116, and in Mann et al, 2011, p. 25. It may be useful to run through a few cases.

Let (μ^*, ν^*) be an equilibrium. Let S_I^* be the set of pure strategies of player I in the support of μ^*, i.e. $\sigma \in S_I^*$ if and only if $\mu^*(\sigma) > 0$. Given that (μ^*, ν^*) is an equilibrium pair we have

(a) for every $\sigma \in S_I$: $U_I(\sigma, \nu^*) \leq U_I(\mu^*, \nu^*)$.

We have also seen that

(b) $U_I(\mu^*, \nu^*)$ may be presented as

$$U_I(\mu^*, \nu^*) = \sum_{\sigma \in S_I} \mu^*(\sigma) \cdot U(\sigma, \nu^*)$$

But we know that for the strategies σ outside the support of μ^*, $\mu^*(\sigma) = 0$. Hence we can take $U_I(\mu^*, \nu^*)$ to be

$$U_I(\mu^*, \nu^*) = \sum_{\sigma \in S_I^*} \mu^*(\sigma) \cdot U_I(\sigma, \nu^*)$$

Now given that $\mu^*(\sigma) > 0$, it cannot be that for every $\sigma \in S_I^*$, $U_I(\sigma, \nu^*)$ is strictly less than $U_I(\mu^*, \nu^*)$, for in that case sum $\sum_{\sigma \in S_I^*} \mu^*(\sigma) U_I(\sigma, \nu^*)$ would be strictly less than $U_I(\mu^*, \nu^*)$, which contradicts (b). So for some $\sigma \in S_I^*$, we must have $U_I(\sigma, \nu^*) = U_I(\mu^*, \nu^*)$. Now, in order to derive a contradiction, suppose that for some $\sigma' \in S_I^*$,

$$U_I(\sigma', \nu^*) < U_I(\mu^*, \nu^*).$$

Then

$$\begin{aligned} U_I(\mu^*, \nu^*) &= \textstyle\sum_{\sigma \in S_I^*} \mu^*(\sigma) U_I(\sigma, \nu^*) \\ &= \mu^*(\sigma') U(\sigma', \nu^*) + \textstyle\sum_{\sigma \in S_I^* - \{\sigma'\}} \mu^*(\sigma) U_I(\sigma, \nu*) \\ &< \mu^*(\sigma^*) U_I(\sigma^*, \nu^*) + \textstyle\sum_{\sigma \in S_I^* - \{\sigma'\}} U_I(\sigma, \nu*). \end{aligned}$$

In other words, player I could improve his expected utility by playing σ^* against σ'. But this contradicts the fact that (μ^*, ν^*) is an equilibrium. We have shown that for every $\sigma \in S_I^*$,

$$U_I(\sigma, \nu^*) = U_I(\mu^*, \nu^*).$$

The proof of (2) is similar. The proofs of (3) and (4) are immediate. For the converse, suppose that clauses (1)-(4) obtain. The sets

$$S_I^* = \{\sigma \in S_I : \mu^*(\sigma) > 0\}$$

and

$$\overline{S_I^*} = \{\sigma \in S_I : \mu^*(\sigma) = 0\}$$

are disjoint and $S_I^* \cup \overline{S_I^*} = S_I$. Then for any $\mu \in \Delta(S_I)$:

$$\begin{aligned} U_I(\mu, \nu^*) &= \textstyle\sum_{\sigma \in S_I^*} \mu(\sigma) U_I(\sigma, \nu^*) + \sum_{\sigma \in \overline{S_I^*}} \mu(\sigma) U_I(\sigma, \nu^*) \\ &\leq \textstyle\sum_{\sigma \in S_I^*} \mu(\sigma) U_I(\sigma^*, \nu^*) + \sum_{\sigma \in \overline{S_I^*}} \mu(\sigma) U_I(\sigma^*, \nu^*) \\ &= U_I(\sigma^*, \nu^*) \textstyle\sum \mu(\sigma) \\ &= U_I(\sigma^*, \nu^*). \end{aligned}$$

This establishes condition (i) of the definition of a Nash equilibrium. Condition (ii) is established in a similar way. □

The above theorem is important for it shows that if the mixed strategy profile (μ, ν) is an equilibrium and σ is any pure strategy in the support of μ then also (σ, ν) is an equilibrium. And if τ is any strategy in the support of ν, then also (μ, τ) is an equilibrium. The theorem is known in the literature as the "purification" of a mixed strategy. It implies that *a player with a mixed strategy in equilibrium must be indifferent between all pure strategies chosen with positive probability.* This, in turn, imposes restrictions on the other player's behaviour, which helps us finding a mixed strategy equilibrium. We look at an example.

Example 4.5 Recall the Matching Pennies Game:

	head	*tails*
heads	$(1,0)$	$(0,1)$
tails	$(0,1)$	$(1,0)$

There is no pure strategy equilibrium. By von Neumann's theorem, we know that there is at least one mixed strategy equilibrium. Here is how we find it using the above theorem.

Since there is no pure equilibrium, player I must mix *heads* and *tails*, that is, he must assign to each of them a positive probability. Notice that there is no equilibrium in which one player plays a pure strategy and the other plays a mixed strategy. This can be seen in a straightforward way: if e.g. player I had a pure strategy and player II a mixed strategy, and the strategies were in equilibrium, then player II would be better off by playing *tails* (he would get payoff 1.) This contradicts our assumption that there is no pure strategy equilibrium.

So let p denote the probability that player I plays *heads* (and thus $1-p$ is the probability that he plays *tails*.) Let q denote the probability that player II plays *heads* (and thus $1-q$ is the probability that he plays *tails*). Then, by the theorem, we must have

$$U_I(h,q) = U_I(t,q).$$

This equation imposes a restriction on player II's behaviour represented by q. Given that:

$$U_I(h,q) = q \cdot 1 + (1-q) \cdot 0 = q$$
$$U_I(t,q) = q \cdot 0 + (1-q) \cdot (1) = 1-q$$

it follows that

$$q = 1-q$$

which yields $q = 1/2$.

A similar argument shows that

$$U_{II}(p,h) = U_{II}(p,t)$$

which yields $p = 1/2$.

We have found the equilibrum in mixed strategies (μ, ν) such that

$$\mu(h) = p = \mu(t) = 1/2$$

and

$$\nu(h) = q = \nu(t) = 1/2.$$

This kind of exaples with two rows and two colums are easy. When the number of rows and columns increases, things become much more difficult. We consider two more examples where the above theorem is used, not to find an equilibrium, but to test whether a given strategy profile is an equilibrium or not.

Example 4.6 We return to the Matching Pennies and the Inverted Matching Pennies games but now played on a universe $M = \{1, 2, 3, 4\}$. We apply the above theorem to show that the strategy pair (μ^*, ν^*) such that both μ^* and ν^* are uniform strategies with suport $M = \{1, 2, 3, 4\}$, is a mixed strategy equilibrium in both games:

	τ_1	τ_2	τ_3	τ_4
σ_1	$(1,0)$	$(0,1)$	$(0,1)$	$(0,1)$
σ_2	$(0,1)$	$(1,0)$	$(0,1)$	$(0,1)$
σ_3	$(0,1)$	$(1,0)$	$(1,0)$	$(0,1)$
σ_4	$(0,1)$	$(0,1)$	$(0,1)$	$(1,0)$

	τ_1	τ_2	τ_3	τ_4
σ_1	$(0,1)$	$(1,0)$	$(1,0)$	$(1,0)$
σ_2	$(1,0)$	$(0,1)$	$(1,0)$	$(1,0)$
σ_3	$(1,0)$	$(1,0)$	$(0,1)$	$(1,0)$
σ_4	$(1,0)$	$(1,0)$	$(1,0)$	$(0,1)$

First notice that in the first game $U_I(\mu^*, \nu^*) = 1/4$ and thus $U_{II}(\mu^*, \nu^*) = 3/4$. Notice that conditions (3) and (4) of the theorem are trivially satisfied. For (1), let $\sigma \in S_I$. Recall that:

$$U_I(\sigma, \nu^*) = \sum_{\tau \in S_{II}} \nu^*(\tau) \cdot u_I(\sigma, \tau).$$

Then:

$$\begin{aligned} \textstyle\sum_{\tau \in S_{II}} \nu^*(\tau) u_I \cdot (\sigma, \tau) &= (1/4 \cdot 1) + (1/4 \cdot 0) + (1/4 \cdot 0) + (1/4 \cdot 0) \\ &= 1/4 \\ &= U_I(\mu^*, \nu^*). \end{aligned}$$

A similar argument establishes (ii). The argument for the second game is similar. Notice that the left side game is an extended Matching Pennies game in which the players choose among four individuals (instead of two). The right side strategic game is an inverted Matching Pennies: the column player wins if the choices match and the first player wins otherwise.

Example 4.7 (Mann et al, 2011) We consider the following 2-player finite win-lose strategic game:

	τ_1	τ_2	τ_3	τ_4
σ_1	$(1,0)$	$(0,1)$	$(0,1)$	$(0,1)$
σ_2	$(0,1)$	$(1,0)$	$(0,1)$	$(0,1)$
σ_3	$(0,1)$	$(1,0)$	$(1,0)$	$(0,1)$
σ_4	$(0,1)$	$(0,1)$	$(0,1)$	$(1,0)$

Consider the pair of mixed strategies (μ^*, ν^*) where μ^* is defined by:

$$\mu^*(\sigma_1) = \mu^*(\sigma_2) = \mu^*(\sigma_2) = 1/5$$

and

$$\mu^*(\sigma_4) = 2/5.$$

The strategy ν^* is defined by:

$$\mu^*(\tau_1) = \mu^*(\tau_2) = \mu^*(\tau_3) = 1/5$$

and

$$\mu^*(\tau_4) = {}^2\!/\!_5.$$

We first compute $U_I(\mu^*, \nu^*)$ and $U_{II}(\mu^*, \nu^*)$. There are 7 outcomes with non-zero payoffs for player I, six of which are chosen with probability $1/5 \times 1/5 = 1/25$ and one with probability $2/5 \times 2/5 = 4/25$. Thus the expected utility of player I is

$$U_I(\mu^*, \nu^*) = 6 \cdot 1/25 + 4/25 = 10/25 = 2/5.$$

Hence $U_{II}(\mu^*, \nu^*) = 3/5$. We now verify clauses (1)-(4) of the Proposition. Clauses 3 and 4 are trivially satisfied. For clause (1), we compute $U_I(\sigma_i, \nu^*)$ for $i = 1, ..., 4$. Take $i = 1$. Recall that to compute $U_I(\sigma_1, \nu^*)$ we look a the sum:

$$U_I(\sigma_1, \nu^*) = \sum_{j=1}^{4} \nu^*(\tau_j) u_I(\sigma_1, \tau_j) \quad = \quad \nu^*(\tau_1) + \nu^*(\tau_3) = 2/5.$$

The calculations for $U_I(\sigma_2, \nu^*)$ and $U_I(\sigma_3, \nu^*)$ are similar and yield the same result. Finally:

$$U_I(\sigma_4, \nu^*) = \sum_{j=1}^{4} \nu^*(\tau_j) u(\sigma_4, \tau_j) = \nu^*(\tau_4) = 2/5.$$

Clause (2) follows automatically. We conclude on the basis of the theorem that (μ^*, ν^*) is an equilibrium.

Chapter 5

Signaling games

5.1 The Stag Hunt and the Prisoner's Dilemma

The Stag Hunt game has become the prototype of a social contract. It is the topic of a whole book by Brian Skyrms, *The Stag Hunt and the Evolution of the Social Structure* (2004). Skyrms traces the history of the game back to Rousseau (Discours sur l'Inegalité)

> If it was the matter of hunting a dear, everyone would realize that he must remain faithful to his post; but if a hare happened to pass by within reach of one of them, we cannot doubt that he would have gone off in pursuit of it without scruples.

and notices that the same game reappears in Hume's *Treatise* in his most famous illustration of a convention:

> two men who pull at the oars of a boat, do it by an agreement or convention though they have never given promises to each other. Both men can either row or not row. If both row, they get the outcome that is best for each - just as, in Rousseau's example, when they hunt the stag. If one decides not to row, then it makes no difference if the other does, or does not - they don't get anywhere. (Hume, 1739/1949, p. 490)

The similarity of the two scenarios is obvious: the worst outcome for you is if you row and the other does not, for then you loose your effort for nothing, just as the worse outcome for you in the Stag Hunt is if you hunt stag by yourself. And Skyrms goes on observing that we meet the Stag Hunt again in Hume's meadow draining problem:

> Two neighbours may agree to drain a meadow which they possess in common because it is easy for them to know each other's mind, and each may perceive that an immediate consequence of failing in his part is the abandonment of the whole project. But it is difficult,

> and indeed impossible that a thousand of persons should agree on such an action. (Idem, p. 538)

Here is its description of the Stag Hunt in one of the standard textbooks on game theory:

> Each of a group of hunters has two options: she may remain attentive to the pursuit of a stag, or she may catch a hare. If all hunters pursue the stag, they catch it and share it equally; if any of the hunters devotes her energy to catching a hare, the stag escapes, and the hare belongs to the defecting hunter alone. Each hunter prefers a share of the stag to a hare. [Osborne 2003, p. 20]

It is customary to represent the game as a 3-player strategic game:

$III = stag$

		II: *stag*	II: *hare*
I	*stag*	(2, 2, 2)	(0, 1, 0)
	hare	(1, 0, 0)	(1, 1, 0)

$III = hare$

	stag	*hare*
stag	(0, 0, 1)	(0, 1, 1)
hare	(1, 0, 1)	(1, 1, 1)

There are two equilibria in the game: (*stag,stag,stag*) and (*hare,hare,hare*). We shall often resort to the 2-player version of the game. Skyrms draws an interesting comparison between the Stag Hunt (SH) and the Prisoner's Dilemma (PD). Here it is.

We already noticed that the PD game can be solved in a straightforward way: each player has a dominant strategy. We also pointed out, however, that there is a tension between the solution of the game, based simply on the rationality of the players, and their mutual benefit. This is consistent with the fact that the solution of the game does not depend on a player's belief about what the other player might do. In the PD game, this kind of solution leads to the players' settling for less.

In the SH game, on the other side, there is no dominant strategy for either player. We have seen that this is a typical situation in cooperation games. There are, instead, two Nash equilibria: $(stag, stag)$ and $(hare, hare)$. This means that it is best to hunt hare if the other hunts hare, and it is best to hunt stag if the other hunts stag. But there is a difference between the two equilibria that Skyrms describes in the following way:

> A player who chooses to hunt stag takes a risk that the other will choose not to cooperate in the Stag Hunt. A player who chooses to

> hunt hare runs no such risk, since his payoff does not depend on the choice of action of the other player, but he foregoes the potential payoff of a successful stag hunt. Here rational players are pulled in one direction by considerations of mutual benefit and in the other by considerations of personal risk. (Skyrms, 2004)

Skyrms calls the (*stag,stag*) a *risk dominant* equilibrium and adds that "That is not to say that rational players could not coordinate at the stag hunt equilibrium that gives them a better payoff; but it is to say that they need a mechanism of trust to do so".

Skyrms (2004) sees Lewis signaling games as a variant of the Stag Hunt. I will follow his analysis later on but before that I will consider shortly the Finnish logician Stenius' *Garden game* that was a source of inspiration for Lewis.

5.2 Stenius: The Garden Game

One traditional view in philosophy of language is that without rules of usage common to the the speaker and the listener, communication would be impossible. According to it every linguistic expression has a meaning that is determined by the rules for its correct use. Erik Stenius gave examples of language games (Garden game) that in his conception illustrate the view of language advocated L. Wittgenstein in his later period.

The *Garden Game* is played by a gardener A and his assistant B. The pieces in the game are the letter ’*a*’, ’*b*’, ’*c*’, ’*P*’ and ’*Q*’. In addition there is a flower bed divided into squares. In every square there is a plant. One of the configurations of the flower bed may be like this:

			3rd
			2d
c	*b*	*a*	*1st*

Every day B writes on a piece of paper the letters ’*a*’, ’*b*’, ’*c*’, and to the left of any of these letters, he writes either the letter ’*P*’ or the letter ’*Q*’, according to whether the plant in the square corresponding for that day to the lowercase letter is in flower or not. For instance, in the rectangle in that day the plant next to the path is in flower, whereas the two others are not, B will write: *PaQbQc*.

The teaching of the game is done by simple gestures of approval and disapproval, depending on whether B writes the correct tokens on the piece of paper.

Once the assistant masters the Garden Game, A and B move to play the *Report Game.* A does not need to accompany B any longer to the flowerbed. A now partakes in the game only by receiving the written tokens from B. If B really follows the rules of the game, A can read off certain facts from what B has been written. A small language for communication has been put in place: ’*a*’, ’*b*’, and ’*c*’ are used to denote certain squares, ’*P*’ and ’*Q*’ express certain properties, etc. These symbols have acquired a meaning.

Stenius's games inspired David Lewis who responded to a challenge launched by Quine. Quine's attack targeted the logical empiricists' conception according to which truths of logic are based on conventions. Quine found this analysis circular for he thought that conventions presuppose already the use of language and logic. Lewis set himself the task to provide an analysis of conventions which does not rely on such presuppositions. In his work conventions are approached in terms of Sender-Receiver, signaling games.

5.3 David Lewis: Signaling systems

D. Lewis set himself the task to provide an analysis of conventions which does not rely on logic and meaning. The outcome is his well known *signaling system.*

A signaling problem is a situation which involves a communicator (sender, speaker) (S) and an audience (receiver, hearer) (R). S observes one of several states s which he tries to communicate or "signal" to R, who does not see s. After receiving the signal, R performs one of several alternative actions, called responses. Every situation s has a corresponding response $b(s)$ that S and R agree is the best response to take when s holds.

To model a Lewisian signaling system, one fixes the following elements:

- A set S of situations or states of affairs, a set Σ of signals, and a set R of responses,
- A function $b : S \to R$ which maps each situation to its best response,
- Functions $f : S \to \Sigma$ which represent S's strategies. Each function tells S what signal to choose for every situation in S. (We hope that using the same symbol for both the Sender and the set of situations will not cause any confusion).
- Functions $g : \Sigma \to R$ which represent R's strategies. Each function tells A which action to perform in response to a signal he receives.

We consider for illustration a signaling system with two states, s_1 and s_2, two messages m_1 and m_2 and two actions a_1 and a_2. C and A both know that a_1 is the best action for the state s_1 and a_2 is the best action for the state s_2. In other words, $b(s_i) = a_i$.

S has four strategies:

$$\begin{aligned}
f_1 &: \{(s_1, m_1), (s_2, m_2)\} \\
f_2 &: \{(s_1, m_2), (s_2, m_1)\} \\
f_3 &: \{(s_1, m_1), (s_2, m_1)\} \\
f_4 &: \{(s_1, m_2), (s_2, m_2)\}
\end{aligned}$$

Likewise, A has four strategies:

$$\begin{aligned}
g_1 &: \{(m_1, a_1), (m_2, a_2)\} \\
g_2 &: \{(m_1, a_2), (m_2, a_1)\} \\
g_3 &: \{(m_1, a_1), (m_2, a_1)\} \\
g_4 &: \{(m_1, a_2), (m_2, a_2)\}
\end{aligned}$$

Lewis's signaling systems have been modelled in two ways: as extensive games; and as strategic games.

5.4 Signaling systems as extensive games with incomplete information

One way to model signaling systems is grounded in Harsanyi's work in the sixties (Harsanyi, 1967). He introduces an external uncertainty about some of the properties of the players or about the rules of the game which consists in supposing that certain parameters, (plays, payoffs, behaviour) of the extensive games with perfect information can take random values: the range of values is common knowledge, but this is not the case about the values that are effectively realized. The "external" uncertainty, which leads to games of *incomplete information*, stands in contrast to internal uncertainty about what the other players will do, which is associated with games of *imperfect information*. (See e.g. Guerrien, 2002, chapter V).

In Harsanyi's framework players have types, and this has consequences for the possible payoffs. The incomplete information is manifest in one player not knowing the type of the other. One classical application is in the centipede game when a player chooses to continue the game (instead of defecting) and the othe player does not know whether the first player is of type "completely rational" or "partially irrational". Lewis's signaling systems constitute yet another well known application. In this case the Receiver R does not know the types (states) of the Sender S.

In order to deal with situations of this kind, Harsanyi introduces a new player, Nature, whose only task is to specify a probability distribution over the set of type profiles of the players. Nature does not have a strategic behaviour in this case. If the initial game is represented as a Kuhn-tree, then Nature will have the initial move picking a type profile with a certain probability. She then informs each player of his type, but not of the types of the others. There is thus an asymmetry of information, and before taking a decision, every player must evaluate the types of the other players, relying on a probabilistic calculus.

After each player comes to know his type, every player determines the strategy which will maximize his expected utility, taking into account the types that the other players may take and the probability of their realization. In a dynamic (sequential) game of incomplete information, the optimal solution will involve the beliefs of the players about the behaviour of the other players. In order to see whether their choices are correct, the players use Bayes' rule and the relevant notion of equilibrium here is that of Bayesian equilibrium. Below we shall give two examples of such games: The New Comer game and Lewis's signaling games. Before doing that, let us dwell a bit more on the distinction incomplete vs. imperfect information.

5.5 Incomplete vs. Imperfect information

We consider our earlier example of the two enterprises which have to decide whether they will launch the production of a new merchandise. (The treatment follows Guerrioen, 2002, p. 93.) There are two choices: p (produce) and np (does not produce). The payoffs of the game are described in the matrix below:

	p	np
p	$(-3,-2)$	$(10,0)$
np	$(0,8)$	$(0,0)$

We will first present this game as an extensive game with imperfect information where first A has the first choice between producing p and not producing, np, after which B has the same choices. B however does not now the choice made by player A, i.e. the two histories p and np are indistinguishable for him. The game tree is shown in figure 5.1

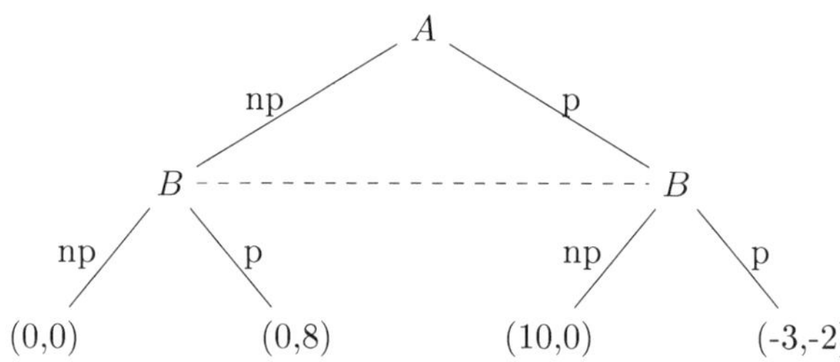

Figure 5.1: The game tree for the game of competing enterprises

We first reproduce B's reasoning. If he finds himself in the upper node and he attributes the probability p_B to p (hence $1 - p_B$ to np), then his expected utility at this node is

$$-2p_B + 0(1 - p_B) = -2p_B.$$

If B finds himself in the lower node, his expected utility is

$$8p_B + 0(1 - p_B) = 8p_B.$$

So if player B believes that A chooses p with probability p_A (and thus chooses np with probability $1 - p_A$), then B's overall expected utility is

$$p_A(-2p_B) + (1 - p_A)8p_B = (8 - 10p_A)p_B.$$

For B the situation is as follows:

- If $8 - 10p_A > 0$, then in order to maximize his expected utility, B would choose $p_B = 1$. If $8 - 10p_A < 0$ then the choice of B must be $p_B = 0$. Finally, if $8 - 10p_A = 0$, then $B's$ choice is arbitrary.

It should be clear that the the choice of B depends on his anticipations of the choice of A.

The reasoning of A is similar:

- If he attributes probability p_A to p (and $1 - p_A$ to np), and he believes that B will choose p with probability p_B then his expected uility is

$$p_A p_B(-3) + p_A(1 - p_B)10 = p_A(10 - 13p_B).$$

If $10 - 13p_B > 0$, then A must choose $p_A = 1$, and if $10 - 13p_B < 0$, then A must choose $p_A = 0$. Finally, if $10 - 13p_B = 0$, then A's choice is arbitrary.

Recall our earlier discussion about rationalization and equilibria in strategic games. We pointed out there that there is equilibrium when the players play their best response to the other player's strategy and in addition this is based on the players correctly predicted what the other will do. The situation is the same here. There is equilibrium if each of the players anticipates correctly what the other will do in such a way that each maximizes his expected utilities. This happens when:

- $p_A = 1$ (player A produces) and $p_B = 0$ (player B does not produce) In this case A gets payoff 10 and B gets payoff 0.
- $p_A = 0$ and $p_B = 1$; In this case A gets payoff 0 and B gets payoff 8.
- $p_A = 4/5$ and $p_B = 10/13$; In this case A gets payoff 0 and B gets payoff 0.

We recover the two equilibria in pure strategies and the mixed strategy equilibrium in our earlier analysis. But now the whole setting is described differently. The extensive form of the game shows better the conditional character of the choices and the beliefs of the players.

We can apply the same reasoning to the Matching Pennies game presented now as an *extensive game of imperfect information*. Recall that player II moves first choosing Heads or Tails after which player I does the same without seeing the choice made by player II earlier. The extensive form of the game is exactly like in the previous example, (with player II being now A and player I being B), except that the payoffs are different: player I gets payoff 1 whenever his choice matches the choice of player II.

As in the previous game, we first put ourselves in the shoes of player I.

- If he is in the upper node, and attributes probability p_I to Heads, then his expected utility at this node is

  $$1p_I + 0(1 - p_I) = p_I.$$

  And if he is at the lower node, then his expected utility at this node is, by a similar reasoning, $(1 - p_I)$. If in addition he attributes probability p_I to the choice of Heads of player I, then his expected utility is:

  $$p_{II}p_I + (1 - p_{II})(1 - p_I) = 2p_Ip_{II} + 1 - p_I \quad p_{II}.$$

  We notice that if $p_{II} = 1/2$ then the expected utility of player I is $1/2$.

A similar reasoning yields the expected utility

$$p_{II}(1-p_I)+(1-p_{II})p_I = p_{II}+p_I-2p_{II}p_I$$

for player II. Thus if p_I is $1/2$, then the expected utility of player II is also $1/2$.

We have recovered the mixed strategy equilibrium in the strategic form of the Matching Pennies game: (σ,τ) where σ and τ are the uniform probability distributions over the set $\{Heads, Tails\}$. (The fact that this pair is an equilibrium follows by some simple calculations). We see once again that formulating the game as an extensive form of imperfect information shows better the conditional character of the choices and the beliefs of the players.

5.6 Sender-Receiver games

(I am indebted to Ratliff's online lecture notes for some part of this material). In this section we look in more details at games of incomplete information using Harsanyi's idea of types introduced earlier. We shall consider only games with two players. We denote by $-i$ the player who is not i. We denote the set of types (or type space) of player i by Θ_i and a type which is a member of this set by $\theta_i \in \Theta_i$. When $\theta_1 \in \Theta_1$ and $\theta_2 \in \Theta_2$ the pair $\theta = (\theta_1, \theta_2)$ is a type profile which is a member of the type space $\Theta = \Theta_1 \times \Theta_2$.

As already mentioned, Nature moves first by picking a type profile θ according to a probability distribution p over $\Theta_1 \times \Theta_2$. We assume that p is objective and thus common knowledge. Given p, each player i can recover the probability of his own type θ_i which is the probability that Nature draws a type profile θ whose i'th component is θ_i that we denote by $p_i(\theta_i)$:

$$p_i(\theta_i) = \sum_{\theta_{-i}\in\Theta_{-i}} p(\theta_i, \theta_{-i}).$$

Although a player does not know the type of his opponent, she forms (subjective) beliefs about it. We represent them by $p(\theta_{-i}/\theta_i)$: the probability that the type of player i is θ_{-i} given that player i observes her own type θ_i. $p(\theta_{-i}/\theta_i)$ is a conditional probability, derived from the objective probability distribution $p(\theta_{-i}, \theta_i)$ which is common knowledge, and the derived subjective probability $p_i(\theta_i)$ of her own type θ_i :

$$p(\theta_{-i}/\theta_i) = \frac{p(\theta_{-i}, \theta_i)}{p_i(\theta_i)}.$$

In the two games we shall consider below, only one player, the Sender, has private information whereas the Receiver does not. We shall take the set of his types to be a singleton.

The game has a sequential nature. After Nature has distributed the types (states) to the Sender, a $\theta-$ Sender first sends a message m, after which the Receiver undertakes a certain action a. The message m, the action a and the type θ of the Sender determine the payoffs of the two players, $u_S(m, a, \theta)$ and $u_R(m, a, \theta)$.

Suppose the Sender's private information is codified by two types $\Theta_S = \{m, n\}$. As mentioned, we take the type space of the Receiver to consist of a singleton, say $\Theta_R = \{r\}$. Then

$$\Theta = \Theta_1 \times \Theta_2 = \{(m, r), (n, r)\}.$$

We assume the types are independent of each other and take a probability distribution p over Θ to be given by

$$p(m, r) = p(n, r) = 1/2.$$

Given p which is common knowledge, each player can recover the probabilities of his types. The probability that the Sender has type m is given by $p(m, r) = 1/2 = p(n, r)$. And the probability that the Receiver has type r is given by

$$p_R(r) = p(m, r) + p(n, r) = 1$$

as expected. The Receiver's subjective belief that the Sender has type m conditional upon the Receivers' knowing his own type r, $p(m/r)$, is given by:

$$p(m/r) = \frac{p(m, r)}{p_R(r)} = \frac{\frac{1}{2}}{1} = 1/2.$$

In a similar way it follows that $p(n/r) = 1/2$. We notice that the Receiver's conditional beliefs about the Sender having a certain type coincide with the Sender's own beliefs about his types. This is due to the fact that the types are independent from each other. This assumption will be maintained in what follows. As a result we shall ignore the type space of the Receiver and denote by Θ the Sender's set of types.

5.6.1 Strategies in Sender-Receiver games

A pure strategy for Sender is a function

$$f : \Theta \to M$$

from the Sender's set of typesΘ to her message space M. For each $\theta \in \Theta$, $f(\theta)$ is a message sent by the type $\theta-$Sender. The Sender can also send mixed (behaviour) strategies which are functions

$$\sigma : \Theta \to \alpha$$

where α is a probability distribution over M. As usual, we shall identify a pure strategy $f : \Theta \to M$ with the mixed strategy which assigns to each $\theta \in \Theta$ the probability distribution α such that for each $m \in M$, $\alpha(m) = 1$ exactly when $m = f(\theta)$ and $\alpha(m) = 0$, otherwise.

Given a mixed strategy σ for the Sender, for any message $m \in M$ we let $\sigma(m/\theta)$ denote the probability according to σ that the type θ sends the message m, i.e. $\sigma(m/\theta) = \sigma(\theta)(m)$.

For a given Sender strategy σ we shall be interested only in the set $M^+(\sigma)$ of on-the-path messages, that is, the messages which are sent by σ with positive probability.

A pure strategy for the Receiver is a function

$$g : M \to A$$

from the set of messages to her action space A. A mixed strategy for the Receiver is a function

$$\rho : M \to \beta.$$

where β is a probability distribution over the Receiver's set of actions A. For any action $a \in A$ we denote by $\rho(a/m)$ the probability, according to the strategy ρ, that the Receiver will do the action a conditional on having observed the message m, i.e., $\rho(a/m) = \rho(m)(a)$.

5.6.2 Utilities and Bayesian equilibria

Consider the case where a type $\theta - Sender$ sends a message $m \in M$ knowing that the Receiver will respond according to his behavior strategy ρ. We introduce three notions.

- The type $\theta-$ Sender's *expected utility* when he sends the message m and the Receiver responds according to the behaviour strategy ρ is given by

$$U_S(m, \rho, \theta) = \sum_{a \in A} \rho(a/m) u_s(m, a, \theta) = \sum_{a \in A} \rho(m)(a) u_s(m, a, \theta).$$

Notice that if, as usually the case, ρ is equivalent to a pure strategy, i.e., $\rho(m)$ is the probability distribution which assigns 1 to exactly one action $b \in A$, then the Sender's expected utility reduces to his payoff in type θ:

$$U_S(m, \rho, \theta) = \rho(m)(b) u_s(m, b, \theta) = u_s(m, b, \theta).$$

- A *pure strategy* f is a *best response* for the Sender to the Receiver's behavior strategy ρ if for every type $\theta \in \Theta$ the message specified by $f(\theta)$ maximizes the Sender's expected utility in type θ. That is, for every type $\theta \in \Theta$:

$$U_S(f(\theta), \rho, \theta) = max_{m \in M} U_S(m, \rho, \theta)$$

- A *behaviour strategy* σ for the Sender is a *best response* to the Receiver behavior strategy ρ if for every type $\theta \in \Theta$, every message m in the support of $\sigma(\theta)$:

$$U_S(m, \rho, \theta)$$

is the maximum in the message space M, i.e.

$$U_S(m, \rho, \theta) = max_{m \in M} U_S(m, \rho, \theta).$$

We now define the same notions for the Receiver. The situation is more complicated given that the Receiver does not know the type of the Sender which sends the message. After receiving a message, he chooses an action, which is optimal given the best beliefs he can have concerning the Sender's type. We have seen that the Receiver has (probabilistic) beliefs p_R concerning the Sender's type θ. Because the Receiver knows the Sender's strategyσ, he might be able to infer something more about the Sender's type and thereby update his prior beliefs p_R. For a message $m \in M$ and a type $\theta \in \Theta$ we denote the Receiver's posterior belief that the Sender's type is θ after Receiver has observed the message m by $p^B(\theta/m)$. There is a correct way to derive $p^B(\theta/m)$ which is given by Bayes' rule:

$$p^B(\theta/m) = \frac{p(\theta) \cdot \sigma(m/\theta)}{\sum_{\theta' \in \Theta} p(\theta')\sigma(m/\theta')}$$

where the notion$\sigma(m/\theta)$ has been defined earlier. Here $p(\theta)\sigma(m/\theta)$ is the probability of the event "type $\theta-$ Sender sends message m" and $\sum_{\theta' \in \Theta} p(\theta')\sigma(m/\theta')$ is the probability of the event "the message m has been sent".

- The Receiver's *expected utility* for action a given the message m is define relatively to the posterior belief p^B:

$$U_R(m, a, p^B) = \sum_{\theta \in \Theta} p^B(\theta/m) \cdot u_R(m, a, \theta)$$

- Given an updated belief system p^B, the Receiver's action a is a *best response* to the message m if the Receiver's expected utility for action a after receiving message m is the maximum in the Receiver's action space A:

$$U_R(m, a, p^B) = max_{a \in A} U_R(m, a, p^B)$$

Consider now a Receiver's pure strategy g.

- g is a *best response* to the Sender's behaviour strategy σ if for all $m \in M^+(\sigma)$, $g(m)$ specifies an action $b \in A$ which is a best response to m given the Receiver's updated belief system p^B:

$$U_R(m, b, p^B) = max_{a \in A} U_R(m, a, p^B)$$

- A Receiver's behavior strategy ρ is a *best response* to the mixed strategy σ if for all $m \in M^+(\sigma)$ every action b in $supp\rho(m)$ is the best response to m given Receiver's Bayes updated beliefs p^B, i.e.,

$$U_R(m, b, p^B) = max_{a \in A} U_R(m, a, \tilde{p}).$$

The updated belief system p^B does not need to be necessarily obtained by using Bayes' rule. However, the notion of Bayesian equilibrium requires it to be so. A *Bayesian equilibrium* of the Sender-Receiver game is a strategy profile(σ, ρ, p^B) such that:

1. The behavior strategy σ for the Sender is a best response to the Receiver behavior strategy ρ

2. The Receiver's behavior strategy ρ is a best response against the Sender's behaviour strategy σ given the posterior belief p^B

3. The posterior belief $p^B$is consistent with Bayes' rule, that is, for every $m \in M^+(\sigma)$, and $\theta \in \Theta$, the value of $p^B(\theta/m)$ is obtained using Bayes' rule.

5.7 First example of a Sender-Receiver game

We consider a variant of the New Comer game discussed earlier but now reformulated as a Sender-Receiver game. The example is from (Guerrien, 2002, chapter V) but our treatment will be different. We model the New Comer game as a Sender-Receiver game where the incomplete information comes from the fact that M (the Receiver) ignores whether NC (the Sender) will come to the market with high costs (hc) or with low ones (lc). The game tree is shown in figure 5.2.

- If NC does not come to the market (nc), then M does not need to do anything (i.e. he will continue to act as before) and will get the payoff 12.

- If NC decides to come to the market (c), then M has to decide whether he accepts or not NC, taking into account that NC comes either in conditions of hc or lc. In the first case, M does not consider NC a serious competitor who will threaten his gains. So his payoff will be only slightly diminished (from 12 to 10). In the second case M is a serious competitor to whom M must accommodate.

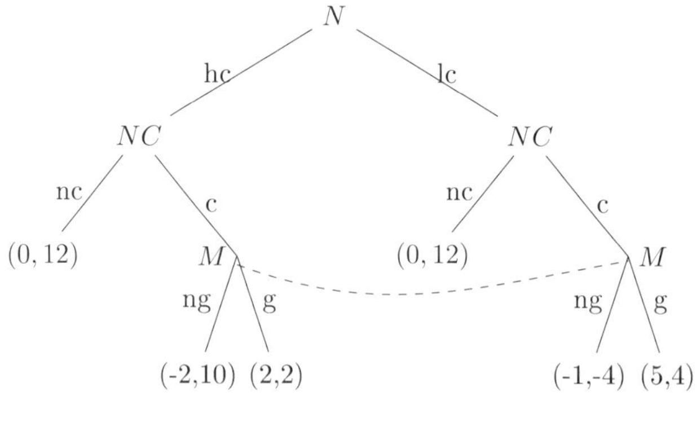

Figure 5.2: The New Comer game with signaling

M knows NC has one of the two types, but not knowing which one, he attributes a probability to this fact in a way which was described earlier.

- Suppose M believes the probabilities are:

$$p(hc) = p(lc) = 1/2.$$

We will show that $(c, c; g; p^B)$ is a Bayesian equilibrium in the New Comer game, where:

- (c, c) is the Sender's pure strategy which sends c in lc and c in hc (equivalently (c, c) is the behaviour strategy σ which maps lc to the degenerated probability distribution which assigns 1 to c and 0 to nc; and σ maps (hc) to the degenerated probability distribution which assigns 1 to c and 0 to nc).)

- g is the Receiver's pure strategy which chooses g after receiving the message c (equivalently, g is the mixed strategy ρ which maps c to the degenerated probability distribution which assigns 1 to g and 0 to ng). Notice that g is not defined on nc.

- p^B is the updated belief system of the Receiver (Monopole) such that $p^B(lc/c) = 1/2$ (and thus $p^B(hc/c) = 1/2$.)

In order to establish condition (1) in the definition of a Bayesian equilibrium, we show that for every type $\theta \in \{lc, hc\}$, ever message in the support of $\sigma(\theta)$ is a best response to ρ in type θ.

For type $\theta = hc$ the only message in the support of $\sigma(\theta)$ is c. Thus we need to show that

$$U_S(c, \rho, hc) \geq U_S(nc, \rho, hc)$$

recalling that

$$U_S(c, \rho, hc) = \sum_{a \in A} \rho(a/c) u_s(c, a, hc).$$

Given thatρ is equivalent to a pure strategy which sends c to a degenerated probability distribution which assigns 1 to g, we know that $U_S(c, \rho, hc)$ reduces to $u_s(c, g, hc) = 2$.

On the other side, if Sender does not come to the market, the game stops and the Sender's expected utility is his payoff. In other words,

$$U_S(nc, \rho, hc) = u_S(nc, hc) = 0.$$

This establishes our claim.

For type $\theta = lc$ the only message in the support of $\sigma(lc)$ is c. We show that

$$U_S(c, \rho, lc) \geq U_S(nc, \rho, lc).$$

Calculations similar to the earlier ones show that

$$U_S(c, \rho, lc) = u_s(c, g, lc) = 5$$

and

$$U_S(nc, \rho, lc) = u_S(nc, lc) = 0$$

which establishes our claim.

Next we show that clause (2) in the definition of Bayesian equilibrium holds. The only on-the-path message $m \in M^+(\sigma)$ is c and the only action in $supp\rho(c)$ is g. We need to show that

$$U_R(c, g, p^B) \geq U_R(c, ng, p^B).$$

We have:

$$\begin{aligned} U_R(c, g, p^B) &= \textstyle\sum_{\theta \in \Theta} p^B(\theta/c) \cdot u_R(c, g, \theta) \\ &= p^B(hc/c) \cdot u_R(c, g, hc) + p^B(lc/c) \cdot u_R(c, g, lc) \\ &= 1/2 \cdot 2 + 1/2 \cdot 4 = 3. \end{aligned}$$

On the other side,

$$\begin{aligned} U_R(c, ng, p^B) &= \textstyle\sum_{\theta \in \Theta} p^B(\theta/c) \cdot u_R(c, ng, \theta) \\ &= p^B(hc/c) \cdot u_R(c, ng, hc) + p^B(lc/c) \cdot u_R(c, ng, lc) \\ &= 1/2 \cdot 10 + 1/2 \cdot (-4) = 3. \end{aligned}$$

We have established the second claim.

Finally we show that the restriction of p^B to on-the-path messages $M^+(\sigma)$ is consistent with Bayes' rule. The only on-the-path message is c. Bayes' rule tells us that

$$p(hc/c) = \frac{p(hc)\sigma(c/hc)}{p(hc)\sigma(c/hc) + p(lc)\sigma(c/lc)} = \frac{1/2 \cdot 1}{1/2 \cdot 1 + 1/2 \cdot 1} = \frac{1}{2} = p^B(hc/c).$$

This establishes our result.

Notice that the Receiver's beliefs are crucial. If the Receiver used a different belief system p^B such that $p^B(hc/c) = 0,77$ (and thus $p^B(lc/c) = 0,23$),

we would have

$$\begin{aligned} U_R(c, g, p^B) &= \textstyle\sum_{\theta \in \Theta} p^B(\theta/c) \cdot u_R(c, g, \theta) \\ &= p^B(hc/c) \cdot u_R(c, g, hc) + p^B(lc/c) \cdot u_R(c, g, lc) \\ &= 0,77 \cdot 2 + 0,23 \cdot 4 = 2,46 \end{aligned}$$

and

$$\begin{aligned} U_R(c, ng, p^B) &= \textstyle\sum_{\theta \in \Theta} p^B(\theta/c) \cdot u_R(c, ng, \theta) \\ &= p^B(hc/c) \cdot u_R(c, ng, hc) + p^B(lc/c) \cdot u_R(c, ng, lc) \\ &= 0,77 \cdot 10 + 0,23 \cdot (-4) = 6,78. \end{aligned}$$

That is, the Receiver's behavior strategy g is not a best response against the Sender's behaviour strategy σ given the new posterior belief system p^B.

5.8 Second example of Sender-Receiver games: Lewis's signaling systems

We now formulate Lewis's signaling systems as Sender-Receiver games with incomplete information. We have already started this task in section 5.3. Nature chooses the type of the Sender S who will then send a message (signal) to the Receiver R. After receiving it, the Receiver will take an apropriate action. There is an asymmetry of information between the Sender S and the Receiver R given that the latter does not know the types (states) of the former. These games are also called "cheap talk games" because, unlike in the New Comer game, the massages do not "matter", i.e., the payoffs of the players do not depend on them. The only thing that matters is whether the two players coordinate with each other to achieve the best action for each type (state). That is, for any state s message m and action a we have:

$$u_C(s,m,a) \quad = u_A(s,m,a) = \begin{cases} 1 & if\ a_j = b(s_i) \\ 0 & otherwise \end{cases}$$

We assume there are two states (types), s_1, s_2, two messages, m_1, m_2, and two actions, a_1, a_2. We assume that the action a_i is the best action for the state s_i. This is primitive information which comes with the specification of the game.

We have already seen that in this extensive game, S will have to choose a message for each state (type), that is, he can follow one of the functions $f_1, ..., f_4$

$$\begin{aligned} f_1 : &\quad \{(s_1, m_1), (s_2, m_2)\} \\ f_2 : &\quad \{(s_1, m_2), (s_2, m_1)\} \\ f_3 : &\quad \{(s_1, m_1), (s_2, m_1)\} \\ f_4 : &\quad \{(s_1, m_2), (s_2, m_2)\} \end{aligned}$$

and the Receiver R has to choose an action for each message, that is, he can follow one of the strategies $g_1, ..., g_4$

$$\begin{aligned} g_1 : &\quad \{(m_1, a_1), (m_2, a_2)\} \\ g_2 : &\quad \{(m_1, a_2), (m_2, a_1)\} \\ g_3 : &\quad \{(m_1, a_1), (m_2, a_1)\} \\ g_4 : &\quad \{(m_1, a_2), (m_2, a_2)\} \end{aligned}$$

We now show that $(f_1; g_1; p^B)$ is a Bayesian equilibrium, where:

- f_1 and g_1 are as above (and can be, equivalently, identified with mixed strategies σ and ρ, in the way described earlier.)
- p^B is the updated belief system for the Receiver such that $p^B(s_1/m_1) = 1$ (and thus $p^B(s_2/m_1) = 0$; and $p^B(s_2/m_2) = 1$ (and thus $p^B(s_1/m_2) = 0$.
- We take the probability distribution over the set of messages to be arbitrary, i.e. $p(s_1) = \gamma$ and thus $p(s_2) = 1 - \gamma$.

In order to establish (1) in the definition of a Bayesian equilibrium, we need to show that for every type $\theta \in \{s_1, s_2\}$, ever message in the support of $f_1(\theta)$ is a best response to g_1 in type θ.

For type $\theta = s_1$ the only message in the support of $f_1(s_1)$ is m_1. Thus we have to show that

$$U_S(m_1, g_1, s_1) \geq U_S(m_2, g_1, s_1).$$

Given that g_1 is equivalent to a pure strategy, we know that $U_S(m_1, g_1, s_1)$ reduces to $u_S(m_1, g_1, s_1) = 1$ which establishes our claim (1 is the maximum value in the game.)

For type $\theta = s_2$ the only message in the support of $f_1(s_2)$ is m_2. We need to show that

$$U_S(m_2, g_1, s_2) \geq U_S(m_1, g_1, s_2).$$

An argument similar to the preceding one establishes that

$$U_S(m_2, g_1, s_2) = u_S(m_2, g_1, s_2) = 1$$

and we are done.

Next we show that (2) in the definition of a Bayesian equilibrium holds. The set of on-the-path messages $M^+(f_1)$ is $\{m_1, m_2\}$. The only action in $suppg_1(m_1)$ is a_1; and the only action in$suppg_1(m_2)$ is a_2.

So we have to show that a_1 is a best response to m_1 given the updated belief system p^B:

$$U_R(m_1, a_1, p^B) \geq U_R(m_1, a_2, p^B).$$

and a_2 is a best response to m_2 given the updated belief system p^B :

$$U_R(m_2, a_2, p^B) \geq U_R(m_2, a_1, p^B).$$

We have

$$\begin{aligned} U_R(m_1, a_1, p^B) &= \textstyle\sum_{\theta \in \Theta} p^B(\theta/m_1) \cdot u_R(m_1, a_1, \theta) \\ &= p^B(s_1/m_1) \cdot u_R(m_1, a_1, s_1) + p^B(s_2/m_1) \cdot u_R(m_1, a_1, s_2) \\ &= 1 \cdot 1 + 0 = 1 \end{aligned}$$

which establishes the first claim. The second claim is similar.

Finally we show that the restriction of p^B to on the path messages m_1 and m_2 is consistent with Bayes' rule. Bayes' rule tells us that

$$p(s_1/m_1) = \frac{p(s_1)\sigma(m_1/s_1)}{p(s_1)\sigma(m_1/s_1) + p(s_2)\sigma(m_1/s_2)} = \frac{\gamma \cdot 1}{\gamma \cdot 1 + (1-\gamma) \cdot 0} = 1.$$

This establishes our result. Notice that the Receiver's posterior beliefs identifies correctly the types of the Sender from the observed messages:

$$p^B(s_1/m_1) = p^B(s_2/m_2) = 1.$$

In connection to this fact, we observe that the probability distribution over the set of states was arbitrary in our example. This means that $\gamma = p(s_1)$ could also be striclty less than 1. In that was case, the posterior belief $p(s_1/m_1) = 1 > \gamma$ effectively updates the Receiver's prior probability γ. The Bayesian equilibrium $(f_1; g_1; p^B)$ is called a *separating equilibrium*.

Let us show that $(f_3; g_3; p^B)$ is not Bayesian equilibrium, where:

- f_3 and g_3 are as above, i.e. $f_3(s_1) = f_3(s_2) = m_1$ and $g_3(m_1) = g_3(m_2) = a_1$. ($f_3$ and g_3 will be identified, as usual, with the correpsonding mixed strategies σ and respectively ρ.)
- p^B is the updated belief system such that $p^B(s_1/m_1) = 1/3$ (and thus $p^B(s_2/m_1) = 2/3$); $p^B(s_2/m_2)$ and $p^B(s_2/m_2)$ do not matter because the message m_2 is never sent.
- We take the probability distribution over the set of messages to be given by $p(s_1) = 1/3$ and $p(s_2) = 2/3$.

Easy calculations show that p^B is consistent with Bayes' rule.

We show that (2) does not hold. The only on the path message is m_1 and the only action in the support of $\rho(m_1)$ is a_1. We show that

$$U_R(m_1, a_1, p^B) \not\geq U_R(m_1, a_2, p^B).$$

We have

$$\begin{aligned} U_R(m_1, a_1, p^B) &= p^B(s_1/m_1)u_R(m_1, a_1, s_1) + p^B(s_2/m_1)u_R(m_1, a_1, s_2) \\ &= 1/3 \cdot 1 + 2/3 \cdot 0 = 1/3. \end{aligned}$$

On the other side

$$\begin{aligned} U_R(m_1, a_2, p^B) &= p^B(s_1/m_1)u_R(m_1, a_2, s_1) + p^B(s_2/m_1)u_R(m_1, a_2, s_2) \\ &= 1/3 \cdot 0 + 2/3 \cdot 1 = 2/3. \end{aligned}$$

Actually if we take the probability distribution over the set of messages to be given by $p(s_1) = 1/3$ and $p(s_2) = 2/3$ we end up with the following Bayesian equilibria

$$(f_1; g_1; p_1^B), (f_2; g_2; p_2^B), (f_3; g_4; p_3^B), (f_4; g_4; p_4^B)$$

where p_i^B is the corresponding belief system obtained using Bayes' rule.

5.9 Lewis' signaling systems: strategic cheap talk games

Lewis' s signalling games may also be formulated as strategic games. In this case we assume a common prior probability function p over the set of states such that $p(s_1)+p(s_2) = 1$. This function is known to both players. Unlike in the dynamic Sender-Receiver games of the previous section, which had a sequential nature, in this case S and R pick up a strategy function independently of the situation in which they find themselves. The set of strategies available to each of them has been listed in an earlier section. For each player i, the payoff $u_i(s, f_i, g_j)$ is $u_p(s, g_j(f_i(s)))$. That is, player i gets payoff 1 if he succeeds in "coordinating" with the other player for the state s and 0 otherwise. Once player S plays one of his strategy functions and player R does the same, the expected utility $U_p(f, g)$ of the player p for the strategy pair (f_i, g_j) is the weighted sum of the actual payoffs for each state:

$$U_p(f_i, g_j) \quad = \quad \sum_{s \in S} \nu(s) u_p(s, f_i, g_j)$$

Suppose p is a uniform probability distribution, that is, $p(s_1) = p(s_2) = 1/2$, then the resulting strategic game looks like this:

	g_1	g_2	g_3	g_4
f_1	$(1, 1)$	$(0, 0)$	$(1/2, 1/2)$	$(1/2, 1/2)$
f_2	$(0, 0)$	$(1, 1)$	$(1/2, 1/2)$	$(1/2, 1/2)$
f_3	$(1/2, 1/2)$	$(1/2, 1/2)$	$(1/2, 1/2)$	$(1/2, 1/2)$
f_4	$(1/2, 1/2)$	$(1/2, 1/2)$	$(1/2, 1/2)$	$(1/2, 1/2)$

The game has 6 equilibria:
$(f_1, g_1), (f_2, g_2), (f_3, g_4), (f_4, g_3), (f_3, g_3)$ and (f_4, g_4).

Recall that for a pair of strategies to be a Nash equilibrium it means that none of the players is better off by unilateral deviation. However, for Lewis only the first two equilibria are signaling systems, because they fulfill one additional assumption: each player, by deviating unilaterally from the equilibrium pair, makes both himself and the other player worse off. The other 4 equilibria are what Lewis calls *noncommunicative equilibria*. They lack the properrty just mentioned. The analogy between communicative equilibria and noncommunicative equilibria, on one side, and the risk dominant equilibria and equilibria which are not risk dominated in the Stag Hunt game, on the other, should be obvious.

The number of equilibria may be diminished if we chose a non-uniform prior probability p. Here is an example of the strategic game where $p(s_1) = 1/3$ and $p(s_2) = 2/3$:

	g_1	g_2	g_3	g_4
f_1	$(1, 1)$	$(0, 0)$	$(1/3, 1/3)$	$(2/3, 2/3)$
f_2	$(0, 0)$	$(1, 1)$	$(1/3, 1/3)$	$(2/3, 2/3)$
f_3	$(1/3, 1/3)$	$(2/3, 2/3)$	$(1/3, 1/3)$	$(2/3, 2/3)$
f_4	$(2/3, 2/3)$	$(1/3, 1/3)$	$(1/3, 1/3)$	$(2/3, 2/3)$

In this case we recover the 4 equilibria $(f_1, g_1), (f_2, g_2), (f_3, g_4)$ and (f_4, g_4) from the dynamic Sender-Receiver game in the previous section. The only communicative equilibria are $(f_1, g_1), (f_2, g_2)$.

We notice that in the communicative equilibria there is a 1-1 correspondence between states, messages and actions. In these cases it is natural to take the actions of the hearer (Receiver) to interpret the messages of the speaker (Sender). Moreover, we can identify the set of actions with the set of states and take the set of states to be the meanings of the corresponding messages, i.e. what the speaker intends to communicate.

The standard example of a signaling problem is that of a driver who is trying to back into a parking space. She has an assistant who gets out of the car and stands in a location where she can simultaneously see how much space there is behind the car. The assistant can be seen by the driver. The assistant wishes to communicate whether there is enough space behind the car for the driver to continue to back up. She has two signals at her disposal: she can stand palms facing in or palms facing out. The driver has two possible responses: she can back up or she can stop.

There are two signaling systems. The assistant can stand palms facing in when there is space, and palms facing out when there is no space, and *vice versa*. In the first case, the driver should continue backing up when she sees the assistant stand palms facing in, and back up when the assistant stands palms facing out. In the second case, the driver should stop when he sees the assistant stands palms facing in, and stop when the assistant stands palms facing out. Both systems work equally well in the sense that the composition of the two communicating and responding strategies realize the best response: the driver back up when there is space, and he stops when there is not.

5.10 Conclusions

It seems we have obtained at least a partial answer to the question *what conventions are*: they are communicative equilibria in Lewis' signaling games. In the noncommunicative equilibria no information is transmitted. In all of them the hearer ignores the message transmitted and always plays the same action which gives him the highest expected utility, if the speaker sends the same message in every situation. On the other side, the speaker has no incentive to send different messages in different states if the hearer always performs the same action. (Benz, 2006).

Skyrms (2004, p. 54) points out that we still do not yet have an answer to the question of how *conventions are selected* from the two equilibria in the game. Lewis himself was aware of the fact that the property of equilibria being risk dominant does not suffice for qualifying them as conventions. Risk dominant equilibria explains only why there is an incentive not to unilaterally deviate but if, for instance, you expect me to deviate, then you might believe to be better off by deviating as well. And if I believe you have such beliefs, I might deviate too. In other words, Lewis saw the need to back up the maintenance of conventions by a hierarchy of interrelated beliefs which are common knowledge. Skyrms (p. 54) is somehow sceptical about the formation of such beliefs in the first place and offers interesting examples which show how equilibria are de facto selected without the assumption of common knowledge.

Skyrms considers also the case in which there are more states than signals, say three states s_1, s_2 and s_3 and two messages m_1 and m_2. (Skirms, 2010, p. 107) First of all, we have to settle what is the best action for each state in this case. Suppose a_1 is the best action for state s_1 and s_3 and a_2 is the best action for state s_2:

	a_1	a_2
s_1	$(1,1)$	$(0,0)$
s_2	$(0,0)$	$(1,1)$
s_3	$(1,1)$	$(0,0)$

In this case the strategy f by the Sender

$$\begin{array}{lcl} s_1 & \rightarrow & m_1 \\ s_2 & \rightarrow & m_2 \\ s_3 & \rightarrow & m_1 \end{array}$$

and the strategy g by the Receiver:

$$\begin{array}{lcl} m_1 & \rightarrow & a_1 \\ m_2 & \rightarrow & a_2 \\ m_3 & \rightarrow & a_1 \end{array}$$

with the obvious payoffs

$$\begin{array}{lclcl} u_1(s_1, f, g) & = & u_2(s_1, f, g) & = & 1 \\ u_1(s_2, f, g) & = & u_2(s_1, f, g) & = & 1 \\ u_1(s_3, f, g) & = & u_2(s_1, f, g) & = & 1 \end{array}$$

can be shown to be an equilibrium. There are other equilibria but only (f, g) and its appropriate permutations are communicative equilibria. (f, g) may be thought of as representing disjunctive information

$$\begin{array}{ccccc} s_1 & & & & \\ & \searrow & m_1 & \rightarrow & a_1 \\ & \nearrow & & & \\ s_3 & & & & \\ s_2 & \rightarrow & m_2 & \rightarrow & a_2 \end{array}$$

in which the signal m_1 carries both the information s_1 and s_3. Thus we will take m_1 to "mean" either s_1 or s_3 and m_2 to mean s_2.

Chapter 6

Game-theoretical semantics

6.1 Introduction

Evaluation games for first-order logic have arisen from the work of Henkin in the 50's and the work of Hintikka in the 70'. Their purpose is to give a game-theoretical characterization of truth and falsity of first-order sentences. The driving force here are the standard quantifiers, the universal "for every x", and the existential "for some x". Their scopal dependencies and independencies in a first-order sentence are recast in terms of the strategic interaction of two players in a game of perfect information, Eloise and Abelard. For instance, in the game associated with the first-order sentence $\forall x \exists y B(x, y)$, the universal player (Abelard) and the existential player (Eloise) choose individuals from the underlying universe of discourse to be the values of the variable x and y, respectively. The order of the choices and the information available to the players are indicated by the syntax of the sentence. Roughly, relatively to the underlying model, the sentence is true if and only if there is a method (strategy) for Eloise that produces a choice b for every choice a by Abelard so that $B(a, b)$ holds in the model. And the sentence is false if there is a strategy for Abelard which yields an individual a for every possible choice b by Eloise, such that $B(a, b)$ does not hold.

When the dependencies and independencies between quantifiers become more complex, the interaction between the two players become less trivial. For instance, in the game associated with the first-order sentence

$$\forall x \exists y \forall x' \exists y' Q(x, x', y, y')$$

and a model M which fixes the extension of the relation Q, a strategy for Eloise will have first to give her a choice b for every individual a chosen by Abelard, and, second, another choice b' for every individuals a and a' chosen by him. But for the second choice, the strategy may use information encoded in her first choice for b. Game theoretical semantics connects in a nice way:

- the notion of truth and falsity with the existence of strategies in semantical games.

- the dependencies and independencies of quantifiers with choice functions.

The last connection is at "the heart of how classical logicians in the twenties viewed the nature of quantification." (Goldfarb 1979).

We shall now make these informal comments more precise. Some of the material in this section follows closely Mann et al (2011).

6.2 First-order languages and models

6.2.1 Languages

A vocabulary is a set $L = \{R, ..., f, ...c, ...\}$ of relation symbols, function symbols, and individual constants. Each relation and function symbol is associated with a natural number, called its arity, in the standard way. (We shall identify individual constants with nullary function symbols). The function symbols in a vocabulary can be combined with other expressions to form terms. Actually the set of L-terms can be defined in a precise way. It is generated by the finite application of the following rules:

- Every variable is an L-term
- Every constant symbol in L is an L-term
- If f is an n-ary function symbol in L and $t_1, ..., t_n$ are L-terms, then $f(t_1, ..., t_n)$ is an L-term.

Thus if the vocabulary is $L = \{+, 0, 1\}$, '0' and '1' are individual constants and thus L-terms, and so are '$+(0,1)$' (that we usually write as '$0+1$') , '$+(x_1, 0)$', '$+(+(0,1),1)$' (that we usually write as '$(0+1)+1$', etc.

Relation symbols can be combined with terms to form (atomic) formulas. Formulas can be combined with logical connectives and quantiers to form (compound) formulas.

Let L be a vocabulary. The first-order formulas generated by the vocabulary L, shortly the L-formulas, are obtained by the finite application of the following rules:

1. If t_1 and t_2 are L-terms, then $t_1 = t_2$ is an L-formula
2. If R is an n-ary relation symbol in L and $t_1, ..., t_n$ are L-terms, then $R(t_1, ..., t_n)$ is an L-formula.
3. If φ is an L-formula, so is $\neg\varphi$.
4. If φ and ψ are L-formulas, so are $(\varphi \vee \psi)$ and $(\varphi \wedge \psi)$.
5. If φ is an $L-$formula, so are $\exists x\varphi$ and $\forall x\varphi$.

Formulas of the form $t_1 = t_2$ or $R(t_1, ..., t_n)$ are called *atomic*. A *literal* is a formula which is either atomic or the negation of an atomic formula. Free and bound occurrences of variables as well as the definition of a sentence are defined in the standard way.

6.2.2 Models

First-order formulas receive an interpretation through models. A model $\mathbb{M}$ for a vocabulary $L = \{R, ..., f, ...c, ...\}$, or an L-model, has the form $\mathbb{M} = (M; R^{\mathbb{M}}, ..., f^{\mathbb{M}}, ..., c^{\mathbb{M}}, ...)$ where M is a non-empty set called the universe of $\mathbb{M}$.

- If R is an n-ary relation symbol, then its interpretation R^M is an n-ary relation on M;
- If f is an n-ary function symbol, then its interpretation f^M is an n-ary function on M; and,
- If c is an individual constant, then its interpretation c^M is an individual (0-place function) in M.

Let $\mathbb{M}$ be an L-model, $\mathbb{M}'$ an L'-model, and$L \subseteq L'$. If $\mathbb{M}$ and $\mathbb{M}'$ have the same universe, and in addition, for every relation symbol R in L, $R^{\mathbb{M}} = R^{\mathbb{M}'}$, for every function symbol f in L, $f^{\mathbb{M}} = f^{\mathbb{M}'}$, then $\mathbb{M}'$ is called an expansion of $\mathbb{M}$ to L'.

6.2.3 Assignments

A model gives an interpretation for the relation and function symbols of a given vocabulary, but not for variables. This job is done by assignments. Let $\mathbb{M}$ be a model. An assignment in $\mathbb{M}$ is a rule for attaching individuals in M to a (finite) set of variables. Thus if we consider the set of variables $\{x_1, x_2\}$, and let $M = \{a, b\}$ then

$$s: \quad x_1 = a \quad x_2 = b$$

is an assignment in $\mathbb{M}$, and

$$s': \quad x_1 = a \quad x_2 = a$$

is another. We prefer to codify assignments by partial functions whose domain is a (finite) set of variables and whose values are individuals in the universe M of $\mathbb{M}$. Thus we represent the two assignments above as

$$s \quad = \quad \{(x_1, a), (x_2, b)\}$$

and

$$s' \quad = \quad \{(x_1, a), (x_2, a)\}$$

respectively. We shall often write $s(x_1) = a$ and $s(x_2) = b$, etc.

If s is an assignment in $\mathbb{M}$, and $a \in M$, then $s(x_i/a)$ is the assignment with domain $dom(s) \cup \{x_i\}$ defined by: $s(x_i/a)(x_j) = s(x_j)$ if $i \neq j$; and $s(x_i/a)(x_j) = a$ if $i = j$. That is, $s' = s(x/a)$ is exactly like s except that $s'(x) = a$. Notice that if $x \in dom(s)$, then the value that s assigns to x is overwritten when we assign it a new value.

For an example, let $\mathbb{M}$ be a model whose universe M consists of two individuals, $M = \{a, b\}$. Let s be the assignment which assigns a to x_1 and b to x_2, i.e. $s = \{(x_1, a), (x_2, b)\}$. Then

$$\begin{array}{lcl} s(x_2/a) & = & \{(x_1, a), (x_2, a)\} \\ s(x_1/a) & = & s \\ s(x_3/a) & = & \{(x_1, a), (x_2, a), (x_3, a)\} \end{array}$$

When the finite set of variables is empty, we let $\varnothing$ denote the empty assignment.

6.2.4 Satisfaction of atomic formulas

Once an assignment s in a model $\mathbb{M}$ is fixed, every term t in the underlying language receives an interpretation $s(t)$ under that assignment:

$$\begin{array}{lcl} s(c) & = & c^{\mathbb{M}} \\ s(f(t_1, ..., t_n) & = & f^{\mathbb{M}}(s(t_1), ..., s(t_n)). \end{array}$$

We are now ready for the definition of satisfaction in a model $\mathbb{M}$ of an atomic formula relative to an assignment s to its free variables:

Definition 6.1 Let L be a vocabulary, M an L-model, and s an assignment in M. Then

$$\begin{array}{rcl} \mathbb{M}, s \models t_1 = t_2 & \text{iff} & s(t_1) = s(t_2) \\ \mathbb{M}, s \models R(t_1, ..., t_n) & \text{iff} & (s(t_1), ..., s(t_n)) \in R^{\mathbb{M}}. \end{array}$$

For an example, let $\mathbb{M} = \{M; R^{\mathbb{M}}, c_1^{\mathbb{M}}, c_2^{\mathbb{M}}\}$ be a model for the language $L = \{c_1, c_2, R\}$ such that $M = \{a, b\}$, $R^{\mathbb{M}} = \{(a, a), (a, b)\}$, $c_1^{\mathbb{M}}$ is a and $c_2^{\mathbb{M}}$ is a. Let s be an assignment such that $s(x) = a$ and $s(y) = b$. Then $(s(x), s(y)) \in R^{\mathbb{M}}$, and from the above definition it follows that $\mathbb{M}, s \models R(x, y)$. Also $(s(c_1), s(c_2)) \in R$, and thus $\mathbb{M}, s \models R(c_1, c_2)$. On the other side, from $(s(y), s(c_1)) \notin R^{\mathbb{M}}$ and the definition, it follows that $\mathbb{M}, s \not\models R(y, c_1)$.

6.3 Semantical games: extensive games of perfect information

A semantical game is played by two players, $\exists$ (Eloise) and $\forall$ (Abelard). They consider a first-order formula φ in negation normal form (i.e. the negation symbol occurs only infront of atomic formulas), a model $\mathbb{M}$, which interprets the vocabulary of φ, and an assignment s in $\mathbb{M}$ which includes the free variables of φ. $\exists$ tries to show that φ is true in $\mathbb{M}$ (relative to the assignment s), and $\forall$ tries to show that φ is false in $\mathbb{M}$. The game starts from the initial position (φ, s), and after each move the players reach a position (ψ, r), where ψ is a subformula of φ and r is an assignment which eventually extends s. Here are the rules of the game:

1. (ψ, r), for ψ a literal. The game stops. If $\mathbb{M}, r \models \psi$ then $\exists$ wins. Otherwise $\forall$ wins. We recall that the notion $\mathbb{M}, r \models \psi$ has been defined in the preceding section.

2. $(\psi \vee \psi', r)$: $\exists$ chooses $\theta \in \{\psi, \psi'\}$ and players move to the position (θ, r).

3. $(\psi \wedge \psi', r)$: $\forall$ chooses $\theta \in \{\psi, \psi'\}$ and players move to the position (θ, r).

4. $(\exists x\psi, r)$: $\exists$ chooses $a \in M$ and players move to the position $(\psi, r(x/a))$.

5. $(\forall x\psi, r)$: $\forall$ chooses $a \in M$ and players move to the position $(\psi, r(x/a))$.

These rules say, essentially, that disjunction and the existential quantifier prompt moves by Eloise whereas conjunction and the universal quantifier prompt moves by Abelard.

We denote the semantical game played with the formula φ, the model $\mathbb{M}$ and the assignment s by $G(\mathbb{M}, s, \varphi)$. The game may be easily reformulated as a finite 2-player, win-lose extensive game with perfect information,

$$G(\mathbb{M}, s, \varphi) = (N, H, Z, P, (u_p)_{p \in N})$$

as in the first part of the book. We prefer to illustrate with an example.

Example. Let φ be the sentence $\forall x \exists y\, x = y$ and $\mathbb{M} = \{a, b\}$. In the initial position $(\varphi, \varnothing)$ ($\varnothing$ is the empty assignment), Abelard can choose either a or b which generates two histories (branches):

$$\begin{aligned} h_a &= (\varphi, \varnothing, (x, a)) \\ h_b &= (\varphi, \varnothing, (x, b)). \end{aligned}$$

Each of these histories is a choice point for Eloise, i.e. $P(h_a) = P(h_b) = \exists$, who can choose either a or b. This further extends the game-tree to four maximal histories (plays) :

$$\begin{aligned} h_{aa} &= (\varphi, \varnothing, (x, a), (y, a)) \\ h_{ab} &= (\varphi, \varnothing, (x, a), (y, b)) \\ h_{ba} &= (\varphi, \varnothing, (x, b), (y, a)) \\ h_{bb} &= (\varphi, \varnothing, (x, b), (y, b)). \end{aligned}$$

Eloise wins the plays h_{aa} and h_{bb}, i.e., $u_\exists(h_{aa}) = u_\exists(h_{bb}) = 1$ given that $\mathbb{M}, \{(x, a), (y, a)\} \models x = y$ and $\mathbb{M}, \{(x, b), (y, b)\} \models x = y$; Abelard wins the plays h_{ab} and h_{ba}, i.e., $u_\forall(h_{ab}) = u_\forall(h_{ba}) = 1$, given that $\mathbb{M}, \{(x, a), (y, b)\} \not\models x = y$ and $\mathbb{M}, \{(x, b), (y, a)\} \not\models x = y$. As we saw, it is customary to represent extensive games in the form of a game-tree: the initial history (φ, s) is the root of the tree and a play proceeds down the branches. The extensive game in our example can be presented in the following tree form as shown in figure 6.1.

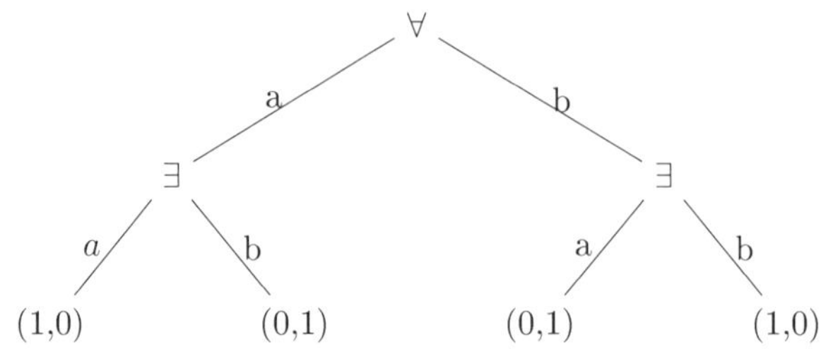

Figure 6.1: The game tree for $G(\mathbb{M}, \emptyset, \forall x \exists y\, x = y)$

6.4 Winning strategies: Game-theoretic truth and falsity

We fix a semantical game $G(\mathbb{M}, s, \varphi)$. When making choices as prescribed by the rules of the game, each player follows a (deterministic) strategy. A strategy for player p in the game $G(\mathbb{M}, s, \varphi)$ is a function σ_p which takes as arguments every (non-maximal) history h where p is to move (i.e. $P(h) = p$) and whose value $\sigma_p(h)$ is the next position to be reached in the game as prescribed by the rules of the game. The strategy σ_p is winning if p wins every maximal history (play) where p follows σ_p.

Notice that in order to determine whether a strategy is winning or not, we need

- to settle the set of maximal plays in which the strategy is followed, and
- to show that all of them are wins for player p.

The notion of following a strategy was analyzed in chapter 2. In the last example, there are four possible strategies for Eloise

$$\begin{array}{lllllll}\sigma_1(h_a) & = & (y,a) & \quad & \sigma_1(h_b) & = & (y,b)\\ \sigma_2(h_a) & = & (y,b) & & \sigma_2(h_b) & = & (y,a)\\ \sigma_3(h_a) & = & (y,a) & & \sigma_3(h_b) & = & (y,a)\\ \sigma_4(h_a) & = & (y,b) & & \sigma_4(h_b) & = & (y,b)\end{array}$$

that we shall sometimes abbreviate as $\sigma_1 = (a,b)$, $\sigma_2 = (b,a)$, $\sigma_3 = (a,a)$, and $\sigma_4 = (b,b)$. Of these, only the strategy σ_1 is winning given that Eloise follows it in the plays h_{aa} and h_{bb} and she wins both of them.

On the other side, Abelard has two possible strategies, τ_a and τ_b,

$$\tau_a(\varphi, \varnothing) \quad = \quad (x,a) \quad \text{and} \quad \tau_a(\varphi, \varnothing) \quad = \quad (x,b)$$

He follows the first in the plays h_{aa} and h_{ab} and the second in the plays h_{ba} and h_{bb}. None of these strategies is winning.

Let $H_\exists$ ($H_\forall$) denote the set of (non-maximal) histories of Z where it is Eloise's (Abelard's turn) turn to move. A player p follows a strategy σ_p in a play (maximal history) h exactly when for each sub-history h' of h such that $h' \in H_p$, $h' \frown \sigma_p(h')$ is a sub-history of h. That is, the "next" member of h after h' is $\sigma_p(h')$. Notice also that when p is Eloise, it follows from the rules of the game that:

- If $P(h') = \exists$ and the last position in h' is $(\theta \vee \theta', r)$, then $\sigma_\exists(h')$ is either θ or θ'.
- If $P(h') = \exists$ and the last position in h' is $(\exists y\psi, r)$, then $\sigma_\exists(h')$ is (y, a) for some $a \in M$.

And when p is Abelard, it follows from the rules of the game that:

- If $P(h') = \forall$ and the last position in h' is $(\theta \wedge \theta', r)$, then $\sigma_\forall(h')$ is either θ or θ'.
- If $P(h') = \forall$ and the last position in h' is $(\forall y\psi, r)$, then $\sigma_\forall(h')$ is (y, a) for some $a \in M$.

Example. Consider the game $G(\mathbb{M}, \varnothing, \varphi)$ where φ is

$$\exists x\forall y(x = y \vee Rxy)$$

and $\mathbb{M} = (\{1,2,3\}, R^M)$ with $R^M = \{(1,2),(2,3)\}$. Here is a winning strategy σ for Eloise. At the beginning of the game Eloise chooses 1

$$\sigma(\oslash, \varphi) = (x, 1).$$

Then, if Abelard chooses 1, Eloise chooses left:

$$\sigma(\oslash, \varphi, (x,1), (y,1)) = x = y.$$

And if Abelard chooses $m = 2, 3$, Eloise chooses right:

$$\sigma(\oslash, \varphi, (x,1), (y,m)) = Rxy.$$

Eloise uses σ in the terminal plays

$$(\oslash, \varphi, (x,1), (y,1), x = y)$$
$$(\oslash, \varphi, (x,1), (y,2), Rxy)$$
$$(\oslash, \varphi, (x,1), (y,2), Rxy)$$

and she wins all of them.

Example. We consider the game $G(\mathbb{M}, s, \varphi)$ where φ is

$$\forall x\exists y(\neg x = y \wedge \neg Rxy)$$

and $\mathbb{M}$ is as in the previous example. A winning strategy for Abelard mimics the winning strategy for Eloise in the previous example. That is, at the beginning of the game, Abelard chooses 1:

$$\tau(\oslash, \varphi) = (x, 1).$$

Then, if Eloise chooses 1, Abelard chooses left:

$$\tau(\oslash, \varphi, (x, 1), (y, 1)) = \neg x = y.$$

And if Eloise chooses $m = 2, 3$, Abelard chooses right:

$$\tau(\oslash, \varphi, (x, 1), (y, m)) = \neg Rxy.$$

Abelard uses τ in the terminal plays

$$\begin{array}{c}(\oslash, \varphi, (x, 1), (y, 1), \neg x = y)\\ (\oslash, \varphi, (x, 1), (y, 2), \neg Rxy)\\ (\oslash, \varphi, (x, 1), (y, 2), \neg Rxy).\end{array}$$

It follows from the definitions that a player may follow his or her strategy in more than one play of the game. However, when Eloise follows one of her strategies, say σ, and Abelard follows one of his strategies, say τ, the result is always a unique play (maximal history) of the game. Thus if in the last example Abelard follows his strategy τ and Eloise follows her strategy σ, $\tau(\oslash, \varphi, (x, 1)) = (y, 1)$, the result is the play

$$(\oslash, \varphi, (x, 1), (y, 1), \neg x = y).$$

We are now ready to define game theoretical truth, $\mathbb{M}, s \models^{+}_{GTS} \varphi$, and game-theoretical falsity, $\mathbb{M}, s \models^{-}_{GTS} \varphi$.

Definition 6.2 Let φ be a first-order L-formula, $\mathbb{M}$ an L-model and s an assignment in M whose domain includes $Free(\varphi)$. Then

- $\mathbb{M}, s \models^{+}_{GTS} \varphi$ iff there is a winning strategy for Eloise in $G(\mathbb{M}, s, \varphi)$
- $\mathbb{M}, s \models^{-}_{GTS} \varphi$ iff there is a winning strategy for Abelard in $G(\mathbb{M}, s, \varphi)$.

6.5 Skolem functions and counter-examples

We prefer to start our considerations with an example.

Example. Consider the sentence $\forall x \exists y \exists z R(x, y, z)$ and a model $\mathbb{M}$. The game is depicted in figure 6.2:

Let σ be a strategy for Eloise in $G(\mathbb{M}, \varnothing, \varphi)$. That is, σ is a function such that for every $a \in M$

$$\sigma(\varphi, \varnothing, (x, a))$$

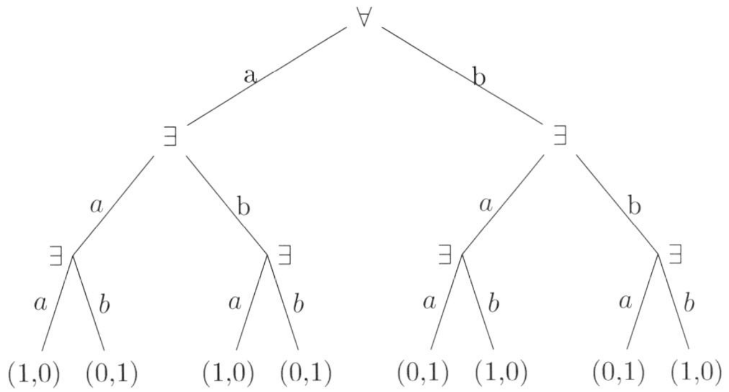

Figure 6.2: The game tree for $\forall x \exists y \exists z R(x, y, z)$ played on $\mathbb{M}$

is a value for y and for every $b \in M$

$$\sigma(\varphi, \varnothing, (x, a), (y, b))$$

is a value for z. Notice that for every $a \in M$ there is exactly one b such that

$$\sigma(\varphi, \varnothing, (x, a)) = (y, b)$$

and for every $a, d \in M$ there is exactly one $e \in M$ such that

$$\sigma(\varphi, \varnothing, (x, a), (y, d)) = (z, e).$$

Both claims follow from the fact that the strategy σ is a function.

We can split σ into two functions $f : M \to M$ and $g : M^2 \to M$ which represent the "local" strategies of Eloise, one for each of her moves. We let f be the function corresponding to $\exists y$ such that for every $a \in M$

$$\sigma(\varphi, \varnothing, (x, a)) = (y, f(a)).$$

We let g be the function corresponding to $\exists z$ such that for every $a, b \in M$

$$\sigma(\varphi, \varnothing, (x, a), (y, b)) = (z, g(a, b)).$$

In a play of the game, Abelard chooses first a value a for x; then Eloise chooses a value b for y, followed by a choice of a value e for z. If Eloise follows the strategy σ or alternatively, the pair of functions (f, g), then these choices are given by $f$and g:

$$(\varphi, \varnothing, (x, a), (y, f(a)), (z, g(a, f(a)))).$$

If, in addition, the strategy σ is winning, Eloise must win every play where she follows it, i.e., we must have

$$R^M(a, f(a), g(a, f(a))),$$

for all $a \in M$.

Suppose now there are functions $f : M \to M$ and $g : M^2 \to M$ such that

$$(*)\ R^M(a, f(a), g(a, f(a)))$$

for every $a \in M$. Then we can define a strategy σ for Eloise in the game $G(M, \varnothing, \varphi)$ in a straightforward way. We let

$$\sigma(\varphi, \varnothing, (x, a)) = (y, f(a))$$

and

$$\sigma(\varphi, \varnothing, (x, a), (y, b)) = (z, g(a, b)).$$

It is easy to see that Eloise wins every play

$$(\varphi, \varnothing, (x, a), (y, b), (z, e)$$

in which she uses σ. For in such a play,

$$(y, b) = \sigma(\varphi, \varnothing, (x, a))$$

and

$$(z, e) = \sigma(\varphi, \varnothing, (x, a), (y, b))$$

and the fact that the play is a win for Eloise follows from the condition (*). Thus we have shown that:

- Eloise has a winning strategy in $G(M, \varnothing, \varphi)$ if and only if there are functions $f : M \to M$ and $g : M^2 \to M$ such that for every a in M:

$$R^M(a, f(a), g(a, f(a))).$$

The functions f and g are known as *Skolem functions.*

Strategies for Abelard reduce to individuals of the domain as values for x. Thus we shall index each strategy τ with the individual in M that it picks up, and represent it as τ_a. Then τ_a is the strategy of Abelard such that

$$\tau_a(\varphi_\infty, \varnothing) = (x, a).$$

Abelard follows his strategy τ_a in any play

$$(\varphi, \varnothing, (x, a), (y, b), (z, e))$$

for every $b, e \in M$. If in addition τ_a is winning, then we must have

$$not : R^M(a, b, e)$$

for all $b, e \in M$.

In a similar way it can be shown that

- Abelard has a winning strategy in $G(M, \varnothing, \varphi)$ if and only if there is an (0-place function) individual $a \in M$ such that for every b, e in M: not $R^M(a, b, e)$.

The individual a is called a *counter-example.*

The last example illustrates well the following passage in Hintikka and Kulas (1985), p. 7:

> The truth of S is defined in GTS as the existence of a winning strategy for Myself (the initial verifier). Now Myself's strategy is a function that tells what to do depending on Nature's earlier moves. The part of such a strategy that pertains to (unnegated) existential quantifiers (quantificational moves by Myself) is codified in what are known as Skolem functions. The existence of a winning strategy for Myself can therefore be expressed by a higher-order sentence that asserts the existence of such Skolem functions as codify partly a strategy that wins against any moves by Nature. Thus, the implications of GTS can be studied by means of a translation of first-order languages into higher-order ones ensuing from GTS....

Indeed, as the example shows, the translation of the sentence $\forall x \exists y \exists z R(x, y, z)$ motivated by the GTS approach is

$$\exists f \exists g \forall x R(x, f(x), g(x, f(x))).$$

Given that our semantical games are 2-player win-lose games of perfect information, Zermelo Theorem, ensures us that the principle of bivalence holds:

Theorem 6.3 (Zermelo, 1913) *For every first-order φ and model $\mathbb{M}$ we have: $\mathbb{M} \models^+_{GTS} \varphi$ or $\mathbb{M} \models^-_{GTS} \varphi$.*

6.6 Logical equivalences as recipes for transforming strategies

An advantage of rephrasing truth and falsity in game-theoretical terms is that many of the logical equivalences of propositional logic like

- Commutativity

$$\varphi \wedge \psi \equiv \psi \wedge \varphi$$

- Associativity

$$\varphi \vee (\psi \vee \theta) \equiv (\varphi \vee \psi) \vee \theta$$
$$\varphi \wedge (\psi \wedge \theta) \equiv (\varphi \wedge \psi) \wedge \theta$$

- Absorption

$$\varphi \vee (\varphi \wedge \psi) \equiv \varphi$$
$$\varphi \wedge (\varphi \vee \psi) \equiv \varphi$$

- Distributivity

$$\varphi \vee (\psi \wedge \theta) \equiv (\varphi \vee \psi) \wedge (\varphi \vee \theta)$$
$$\varphi \wedge (\psi \vee \theta) \equiv (\varphi \wedge \psi) \vee (\varphi \wedge \theta)$$

may be interpreted as recipes for converting one winning strategy into another in any underlying model. (For further discussions of this point we refer the reader to Mann et al (2011), which contains detailed proofs of the logical equivalences listed above, and Van Benthem (2014).)

We give a detailed proof of absorption. Suppose that Eloise has a winning strategy σ in $G(\mathbb{M}, s, \varphi)$. Define a winning strategy τ for Eloise in $G(\mathbb{M}, s, (\varphi \vee (\varphi \wedge \psi))$ in the following way:

$$\tau(s, \varphi \vee (\varphi \wedge \psi)) = \varphi$$
$$\tau(s, \varphi \vee (\varphi \wedge \psi), \varphi, ...) = \sigma(s, \varphi, ...)$$

That is, τ is the strategy which tells Eloise to choose left and then mimic σ for the rest of the game. Every terminal history $h' = (s, \varphi \vee (\varphi \wedge \psi), \varphi, ...)$ in the second game in which Eloise follows τ corresponds exactly to one terminal history $h = (s, \varphi, ...)$ in the first game in which Eloise follows σ. Therefore Eloise wins h' if and only if she wins h. But Eloise wins h (given that σ is a winning strategy) and thus she also wins h'. We conclude that τ is a winning strategy.

For the converse, suppose Eloise has a winning strategy τ in $G(\mathbb{M}, s, \varphi \vee (\psi \wedge \theta))$. We define a winning strategy σ for Eloise in $G(\mathbb{M}, s, \varphi)$ as follows. If $\tau(s, \varphi \vee (\varphi \wedge \psi)) = \varphi$, then we let

$$\sigma(s, \varphi, ...) = \tau(s, \varphi \vee (\varphi \wedge \psi), \varphi, ...).$$

And if $\tau(s, \varphi \vee (\varphi \wedge \psi)) = (\varphi \wedge \psi)$, then we let

$$\sigma(s, \varphi, ...) = \tau(s, \varphi \vee (\varphi \wedge \psi), (\varphi \wedge \psi), \varphi ...).$$

In other words, if τ tells Eloise to choose φ, then σ tells her to mimic τ. If τ tells Eloise to choose $(\varphi \wedge \psi)$, then σ tells her to mimic the part of τ she follows when Abelard chooses φ.

We give an informal proof of Distributivity. Suppose Eloise has a winning strategy σ in the game $G(\mathbb{M}, s, (\varphi \vee (\psi \wedge \theta)))$. Suppose the first choice of Eloise according to σ is φ. Define a winning strategy τ for Eloise in $G(\mathbb{M}, s, (\varphi \vee \psi) \wedge (\varphi \vee \theta))$ as follows. If Abelard chooses $\varphi \vee \psi$, then let τ pick up φ and then mimic σ for the rest of the game in φ. If Abelard chooses $\varphi \vee \theta$, then let τ pick up φ and mimic σ for the rest of the game in φ.

Suppose now that the first choice of Eloise according to σ is $(\psi \wedge \theta)$. Then σ must be a winning strategy in both ψ and θ. Define a winning strategy τ for Eloise in $G(\mathbb{M}, s, (\varphi \vee \psi) \wedge (\varphi \vee \theta))$ as follows. If Abelard chooses $\varphi \vee \psi$, then let τ pick up ψ and mimic σ for the rest of the game in ψ. If Abelard chooses $\varphi \vee \theta$, then let τ pick up θ and mimic σ for the rest of the game in θ. The converse is shown in a similar way.

For the converse, suppose that Eloise has a winning strategy σ in the game $G(\mathbb{M}, s, (\varphi \vee \psi) \wedge (\varphi \vee \theta))$. σ must be winning against any choice made by Abelard, i.e. σ wins $(\varphi \vee \psi)$ and it wins $(\varphi \vee \theta)$. Here is a winning strategy τ for Eloise in the game $G(\mathbb{M}, s, \varphi \vee (\psi \wedge \theta))$ σ is winning in both $(\varphi \vee \psi)$ and $(\varphi \vee \theta)$ if and only if σ is winning in φ or σ is winning in both ψ and θ. If the former then let τ choose left (i.e. φ and then copy σ in φ. If the latter then let τ choose right (i.e., $(\psi \wedge \theta)$) and then follow σ in ψ and θ.

The "strategy transformation" perspective may also be adopted for quantifier laws:

$$\begin{aligned} \exists x(\varphi \vee \psi) &\equiv \exists x\varphi \vee \exists x\psi \\ \forall x(\varphi \wedge \psi) &\equiv \forall x\varphi \wedge \forall x\psi \end{aligned}$$

We prove the first one. Suppose Eloise has a winning strategy σ in $G(\mathbb{M}, s, \exists x(\varphi \vee \psi))$ such that

$$\begin{aligned} \sigma(s, \exists x(\varphi \vee \psi)) = (x, a) \\ \sigma(s, \exists x(\varphi \vee \psi, (x, a)) = \chi \end{aligned}$$

where χ is either φ or ψ. We now define a winning strategy τ in the game $G(\mathbb{M}, s, \exists x\varphi \vee \exists x\psi)$ as follows:

$$\begin{aligned} &\tau(s, \exists x\varphi \vee \exists x\psi) = \exists x\chi \\ &\tau(s, \exists x\varphi \vee \exists x\psi, \exists x\chi) = (x, a) \\ &\tau(s, \exists x\varphi \vee \exists x\psi, \exists x\chi, (x, a), ...) = \sigma(s, \exists x(\varphi \vee \psi, (x, a), \chi, ...) \end{aligned}$$

That is, if σ tells Eloise to choose first a as a value of x and then choose e.g., φ, then we let τ to be such that Eloise chooses first $\exists x\varphi$ and then a as a value of x. We observe that in both games after the first two moves, the current assignment is $s(x/a)$ and the current subformula is χ. Thus the game proceeds as in $G(\mathbb{M}, s(x/a), \chi)$. Every terminal history

$$h' \quad = \quad (s, \exists x\varphi \vee \exists x\psi, \exists x\chi, (x, a), ...)$$

of the game $G(\mathbb{M}, s, \exists x\varphi \vee \exists x\psi)$ in which Eloise follows τ corresponds exactly to a terminal history

$$h \quad = \quad (s, \exists x(\varphi \vee \psi), (x, a), \chi, ...)$$

of the game $G(\mathbb{M}, s, \exists x(\varphi \vee \psi))$ in which Eloise follows σ : both histories induce the same assignment and end up with the same atomic formula. Thus Eloise wins h' if and only if she wins h. But she wins h (given that σ is a winning strategy), therefore she also wins h'. Thus τ is a winning strategy.

The proof in the other direction is similar. Notice that it does not matter that we use syntactically the same existentially quantified variable in the same disjunct, i.e. we could state the above quantificational laws as:

$$\begin{aligned} \exists x(\varphi \vee \psi) &\equiv \exists y\varphi \vee \exists z\psi \\ \forall x(\varphi \wedge \psi) &\equiv \forall y\varphi \wedge \forall z\psi. \end{aligned}$$

It turns out to be useful to compare the preceding proof with that of

$$\exists x(\varphi \wedge \psi) \quad \models \quad \exists x\varphi \wedge \exists x\psi.$$

Suppose Eloise has a winning strategy σ in the game $G(\mathbb{M}, s, \exists x(\varphi \wedge \psi))$ such that $\sigma(s, \exists x(\varphi \wedge \psi)) = (x, a)$. We define a strategy τ for Eloise in $G(\mathbb{M}, s, \exists x\varphi \wedge \exists x\psi)$ as follows:

$$\tau(s, \exists x\varphi \wedge \exists x\psi, \exists x\varphi) = (x, a)$$
$$\tau(s, \exists x\varphi \wedge \exists x\psi, \exists x\varphi, (x, a), ...) = \sigma(s, \exists x(\varphi \wedge \psi), (x, a), \varphi, ...)$$

and

$$\tau(s, \exists x\varphi \wedge \exists x\psi, \exists x\psi) = (x, a)$$
$$\tau(s, \exists x\varphi \wedge \exists x\psi, \exists x\psi, (x, a), ...) = \sigma(s, \exists x(\varphi \wedge \psi), (x, a), \psi, ...).$$

That is, τ is the strategies which choose a as a value for x and then mimics σ thereafter.

We observe that in both games, after the players make their first moves, the current assignment is $s(x/a)$ and the current subformula is φ or ψ. In either case, every terminal history h' in $G(\mathbb{M}, s, \exists x\varphi \wedge \exists x\psi)$ in which Eloise follows τ corresponds to a terminal history h in $G(\mathbb{M}, s, \exists x(\varphi \wedge \psi))$ in which Eloise follows σ. Both histories induce the same assignments and terminate with the same atomic subformula. Thus Eloise wins h if and only if Eloise wins h'. But Eloise wins h so she wins h'. Thus τ is a winning strategy in the game $G(\mathbb{M}, s, \exists x\varphi \wedge \exists x\psi)$.

We can now understand better the reason why existential quantifiers distribute over disjunctions but not over conjunctions: Disjunctions prompt moves by Eloise whereas conjunctions prompt moves by Abelard. In the former case, Eloise can plan her choices ahead and choose the value of x that will verify the appropriate disjunct or choose the disjunct first and then choose the value of x. In the second case she is forced to choose the value of x before she knows which conjunct Abelard will choose. In other words, Eloise cannot transform a winning strategy σ for Eloise in the game $G(\mathbb{M}, s, \exists x\varphi \wedge \exists x\psi)$ into a winning strategy in the game $G(\mathbb{M}, s, \exists x(\varphi \wedge \psi))$.

Universal quantifiers distribute over conjunctions but not over disjunction for the same reasons.

A similar argument shows that adjacent existential quantifiers commute and so do adjacent universal quantifiers. On the other side, existential quantifiers do not commute with universal quantifiers. That is, the following hold:

$$\begin{aligned} \exists x \exists y \varphi &\equiv \exists y \exists x \varphi \\ \forall x \forall y \varphi &\equiv \forall y \forall x \varphi \\ \exists x \forall y \varphi &\models \forall y \exists x \varphi \end{aligned}$$

6.7 Game-theoretical negation

There are mainly two ways to think about ordinary sentential negation. On one conception, negation is a form of complementation that operates a partition of the set of all possible models into those that verify p and those that falsify p. It corresponds to the denial of a proposition p as equivalent to making an assertion that holds exactly in those situations in which p fails.

For the other way to look at negation, here is a telling passage:

> The other way of thinking about sentential negation is well exemplified by Ramsey's ... suggestion that the negation of a proposition p be represented by writing p upside down. In this sense negation is a form of reversal or inversion: to deny a given proposition is to say that things are the other way around. Semantically, the same intuition can be found also in Beh-Mann's typographical convention of writing 'T' for truth and '' for falsehood..., or in Hintikka's game-theoretic characterization of negation as swapping of roles between players..., or even in Peirce's rule for "negativing" a graph by cutting through the Phemic sheet and turning over the excised piece.... Extensionally, of course, these two conceptions of negation are equivalent, in that both yield the same truth-functions. This equivalence finds its most obvious representation in the familiar apparatus of truth-tables.... Intensionally, however, the two conceptions are quite distinct. (Varzi and Warglien, 2003, p. 1)

In this section we relax the assumption that negation occurs only in front of atomic formulas. This creates the need for a game-rule handling negation. In game-theoretical semantics the rule is given in terms of the players "switching roles". We thus need roles for the two players.

- At the beginning of the game $\exists$ is the Verifier (V) and $\forall$ is the Falsifier (F). At each history (other than the initial one), we know who is the verifier (V) and who is the falsifier (F) by a simple calculation: if an even number of switches from formulas of the form $\neg\chi$ to formulas of the form χ have taken place, then Eloise is V and Abelard is F; otherwise Eloise is F and Abelard is V.

The rules of the game $G(\mathbb{M}, s, \varphi)$ are now restated as:

- $(P(t_1, ..., t_n), r)$: The game stops. If r satisfies $P(t_1, ..., t_n)$, then (the player who is) V wins. Otherwise F wins.
- $(\neg\psi, r)$: Players move to (B, r) with roles inversed.
- $(\psi \vee \psi', r)$: V chooses $\theta \in\{\psi,\psi'\}$ and players move to (θ, r).
- $(\psi \wedge \psi', r)$: F chooses $\theta \in \{\psi, \psi'\}$ and players move to (θ, r).
- $(\exists x\psi, r)$: V chooses $a \in M$ and players move to $(\psi, r(x/a))$.
- $(\forall x\psi, r)$: F chooses $a \in M$ and players move to $(\psi, r(x/a))$.

Example. Consider the game $G(\mathbb{M}, \varnothing, \neg\varphi)$ where φ is the sentence $\forall x\exists y x = y$ and $\mathbb{M} = \{a, b\}$. In the initial position $(\varnothing, \neg\varphi)$ Eloise is the verifier and Abelard is the falsifier. The players switch roles, which generates the history:

$$h^*_{\neg} = (\varnothing, \neg\varphi, \varphi)$$

This history prompts a move by the falsifier who is Eloise. She can choose either a or b which generates two histories:

$$\begin{array}{rcl} h_a^* & = & (\varnothing, \neg\varphi, \varphi, (x,a)) \\ h_b^* & = & (\varnothing, \neg\varphi, \varphi, (x,b)) \end{array}$$

Each of them is a choice point for the verifier, Abelard, who can choose either a or b. This results in four maximal histories (plays):

$$\begin{array}{rcl} h_{aa}^* & = & (\varnothing, \neg\varphi, \varphi, (x,a), (y,a)) \\ h_{ab}^* & = & (\varnothing, \neg\varphi, \varphi, (x,a), (y,b)) \\ h_{ba}^* & = & (\varnothing, \neg\varphi, \varphi, (x,b), (y,a)) \\ h_{bb}^* & = & (\varnothing, \neg\varphi, \varphi, (x,b), (y,b)). \end{array}$$

Notice that Abelard wins h_{aa}^* and h_{bb}^* and Eloise wins the remaining two histories. There are four strategies for Abelard

$$\begin{array}{rclcrcl} \tau_1(h_a^*) & = & (y,a) & & \tau_1(h_b^*) & = & (y,b) \\ \tau_2(h_a^*) & = & (y,b) & & \tau_2(h_b^*) & = & (y,a) \\ \tau_3(h_a^*) & = & (y,a) & & \tau_3(h_b^*) & = & (y,a) \\ \tau_4(h_a^*) & = & (y,b) & & \tau_4(h_b^*) & = & (y,b) \end{array}$$

Of these, only the strategy σ_1 is winning: Abelard follows it in the plays h_{aa}^* and h_{bb}^* that she both wins.

It is useful to compare this example to our earlier example $G(\mathbb{M}, \varnothing, \varphi)$. The maximal histories in the two games are in 1-1 correspondence to each other.

The notion of strategy for a player p is defined exactly as before, as are game-theoretical truth (existing of a winning strategy for Eloise) and falsity (existing of a winning strategy for Abelard).

We mention few elementary facts about negation in the game-theoretical setting.

Proposition 6.1 Let φ be a first-order formula, M a model and s an assignment in $\mathbb{M}$ whose domain includes the free variables of φ. Then:

(i) There is a winning strategy for Eloise in $G(\mathbb{M}, s, \neg\varphi)$ if and only if there is a winning strategy of Abelard in $G(\mathbb{M}, s, \varphi)$:

$$\mathbb{M}, s \models^+_{GTS} \neg\varphi \quad \texttt{iff} \quad \mathbb{M}, s \models^-_{GTS} \varphi$$

(ii) There is a winning strategy for Eloise in $G(\mathbb{M}, s, \varphi)$ if and only if there is a winning strategy of Abelard in $G(\mathbb{M}, s, \neg\varphi)$:

$$\mathbb{M}, s \models^+_{GTS} \varphi \quad \texttt{iff} \quad \mathbb{M}, s \models^-_{GTS} \neg\varphi.$$

Proof (i) Assume there is a winning strategy σ for Eloise in $G(\mathbb{M}, s, \neg\varphi)$. We define a strategy τ for Abelard in $G(\mathbb{M}, s, \varphi)$ by:

$$\tau(s, \varphi, ...) \quad = \quad \sigma(s, \neg\varphi, \varphi, ...).$$

That is, τ mimics σ in $G(\mathbb{M}, s, \varphi)$. This is possible, given that for every history $h' = (s, \varphi, ...)$, the moves done by Eloise in h' are done by Abelard in $h = (s, \neg\varphi, \varphi, ...)$. It remains to show that τ is a winning strategy. Let $h' = (s, \varphi, ...)$ be a play (maximal history) in $G(\mathbb{M}, s, \varphi)$ where τ is followed. Then $h = (s, \neg\varphi, \varphi, ...)$ is the corresponding play in $G(\mathbb{M}, s, \neg\varphi)$ where σ is followed. Given that σ is winning, h is a win by Eloise. But given the "role reversal" interpretation of negation, there is one less such reversal in h' compared to h. Thus h' is a win for Abelard and we conclude that τ is a winning strategy for Abelard in $G(\mathbb{M}, s, \varphi)$. The converse of (i) is similar and so is (ii). □

Now it should be obvious that even after introducing roles for the two players, we are still inside the class of finite, 2-player, win-lose extensive games of perfect information. But then Zermelo's theorem still applies: either Eloise or Abelard has a winning strategy in the relevant game. Using Zermelo's theorem one can show that game-theoretical negation is contradictory negation.

Proposition 6.2 For every first-order formula φ model $\mathbb{M}$ and assignment s in M whose domain includes the free variables of φ :

$$\mathbb{M}, s \models^{+}_{GTS} \neg\varphi \quad \text{iff} \quad \mathbb{M}, s \not\models^{+} \varphi \ .$$

PROOF Suppose that $\mathbb{M}, s \models^{+}_{GTS} \neg\varphi$. By the first proposition, Abelard has a winning strategy in $G(\mathbb{M}, s, \varphi)$. But then Eloise cannot have one (recall that $G(\mathbb{M}, s, \varphi)$ is a win-lose game). Conversely, suppose that Eloise does not have a winning strategy in $G(\mathbb{M}, s, \varphi)$. Then by Zermelo's theorem, Abelard has one, i.e. $\mathbb{M}, s \models^{-}_{GTS} \varphi$ and by the previous proposition we have $\mathbb{M}, s \models^{+}_{GTS} \neg\varphi$. □

Proposition 6.3 Let φ be a first-order formula, $\mathbb{M}$ a model and s an assignment in $\mathbb{M}$ whose domain includes the free variables of φ. Then there is a winning strategy for Eloise in $G(\mathbb{M}, s, \neg\neg\varphi)$ if and only if there is a winning strategy for Eloise in $G(\mathbb{M}, s, \varphi)$:

$$\mathbb{M}, s \models^{+}_{GTS} \neg\neg\varphi \quad \texttt{iff} \quad \mathbb{M}, s \models^{+}_{GTS} \varphi.$$

PROOF By the first proposition, Eloise has a winning strategy in $G(\mathbb{M}, s, \neg\neg\varphi)$ if and only if Abelard has a winning strategy in $G(\mathbb{M}, s, \neg\varphi)$ if and only if Eloise has a winning strategy in $G(\mathbb{M}, s, \varphi)$. □

Proposition 6.4 Let φ be a first-order formula, $\mathbb{M}$ a model and s an assignment in $\mathbb{M}$ whose domain includes the free variables of φ. Then:

$$\begin{array}{lrcl}
(i) & \mathbb{M}, s \models^{+}_{GTS} \neg(\varphi \wedge \psi) & \texttt{iff} & \mathbb{M}, s \models^{+}_{GTS} (\neg\varphi \vee \neg\psi) \\
(ii) & \mathbb{M}, s \models^{+}_{GTS} \neg(\varphi \vee \psi) & \texttt{iff} & \mathbb{M}, s \models^{+}_{GTS} (\neg\varphi \wedge \neg\psi) \\
(iii) & \mathbb{M}, s \models^{+}_{GTS} \neg\exists x\varphi & \texttt{iff} & \mathbb{M}, s \models^{+}_{GTS} \forall x\neg\varphi \\
(iv) & \mathbb{M}, s \models^{+}_{GTS} \neg\forall x\varphi & \texttt{iff} & \mathbb{M}, s \models^{+}_{GTS} \exists x\neg\varphi
\end{array}$$

PROOF (i) Let σ be a winning strategy for Eloise in $G(\mathbb{M}, s, \neg(\varphi \wedge \psi))$. We define a winning strategy σ' for Eloise in $G(\mathbb{M}, s, \neg\varphi \vee \neg\psi)$ in the following way:

$$\sigma'(s, \neg\varphi \vee \neg\psi) \quad = \quad \begin{cases} \neg\varphi & if\ \sigma(s, \neg(\varphi \wedge \psi), \varphi \wedge \psi) = \varphi \\ \neg\psi & if\ \sigma(s, \neg(\varphi \wedge \psi), \varphi \wedge \psi) = \psi \end{cases}$$

and

$$\begin{aligned}\sigma'(s, \neg\varphi \vee \neg\psi, \neg\varphi, \varphi, ...) &= \sigma(s, \neg(\varphi \wedge \psi), \varphi \wedge \psi, \varphi, ...)\\ \sigma'(s, \neg\varphi \vee \neg\psi, \neg\psi, \psi ...) &= \sigma(s, \neg(\varphi \wedge \psi), \varphi \wedge \psi, \psi, ...).\end{aligned}$$

This is, again, a "copy cat" strategy. Note that this strategy is well defined. Note that every play $h' = (s, \neg\varphi \vee \neg\psi, \neg\chi, \chi, ...)$ in $G(\mathbb{M}, s, \neg\varphi \vee \neg\psi)$ where Eloise follows σ' corresponds to a play $h = (s, \neg(\varphi\wedge\psi), \varphi\wedge\psi, \chi, ...)$ in $G(\mathbb{M}, s, \neg(\varphi\wedge\psi))$ where Eloise follows σ. Now it is easy to see that the number of role reversals in h and h' are identical and given that σ wins h then σ' wins h'. We conclude that σ' is a winning strategy. For the converse, assume that Eloise has a winning strategy σ in $G(\mathbb{M}, s, \neg\varphi \vee \neg\psi)$. We define a strategy σ' for Eloise in the game $G(\mathbb{M}, s, \neg(\varphi \wedge \psi))$ in the following way:

$$\sigma'(s, \neg(\varphi \wedge \psi), \varphi \wedge \psi) = \begin{cases}\varphi & if\ \sigma(s, \neg\varphi \vee \neg\psi) = \neg\varphi\\ \psi & if\ \sigma(s, \neg\varphi \vee \neg\psi) = \neg\psi\end{cases}$$

and

$$\begin{aligned}\sigma'(s, \neg(\varphi \wedge \psi), \varphi \wedge \psi, \varphi, ...) &= \sigma(s, \neg\varphi \vee \neg\psi, \neg\varphi, \varphi, ...)\\ \sigma'(s, \neg(\varphi \wedge \psi), \varphi \wedge \psi, \psi, ...) &= \sigma(s, \neg\varphi \vee \neg\psi, \neg\psi, \psi, ...).\end{aligned}$$

The proof of (ii) is similar. For (iv) let σ be a winning strategy for Eloise in $G(\mathbb{M}, s, \neg\forall x\varphi)$. We define a winning strategy σ' for Eloise in $G(\mathbb{M}, s, \exists x\neg\varphi)$ in the following way:

$$\sigma'(s, \exists x\neg\varphi) = \sigma(s, \neg\forall x\varphi, \forall x\varphi)$$

Assume now $\sigma'(s, \exists x\neg\varphi) = (x, a)$. Then

$$\sigma'(s, \exists x\neg\varphi, (x, a)...) = \sigma(s, \neg\forall x\varphi, \forall x\varphi, (x, a), ...)$$

Note that every play $h' = (s, \exists x\neg\varphi, (x, a)...)$ where Eloise follows σ' corresponds to a play $h = (s, \neg\forall x\varphi, \forall x\varphi, (x, a), ...)$ where Eloise follows σ. As the number of negations in these plays is identical, then Eloise wins one if and only if she wins the other. But σ is a winning strategy, hence by a reasoning we have rehearsed several times, σ' is a winning strategy. The converse is similar. □

The last Proposition allows us to adjust the position of negation symbols in a first-order formula while preserving its truth-conditions. For instance, we can push every negation symbol as deep into the formula as possible, stopping when we reach an atomic formula. A first-order formula φ is in negation normal form if the only negated subformulas of φ are atomic. By repeated application of this proposition, we can show that every first-order formula is equivalent to a formula in negation normal form, which means that we can delay role-reversals until the end of the semantic game.

There is another property of game-theoretical negation that is worth mentioning: it is dual negation, that is, it travels through the quantifiers and connectives in its scope, changes them into their duals (i.e. it changes $\exists x$, $\forall x$, $\vee$, and $\wedge$ into $\forall x$, $\exists x$, $\wedge$, and $\vee$ respectively) until it reaches an atomic formula. Instead of given a formal proof we prefer an example:

$\mathbb{M}, s \models^{\pm}_{GTS} \neg\exists x\forall y\exists z\forall w R(x, y, z, w)$ iff $\mathbb{M}, s \models^{\pm}_{GTS} \forall x\exists y\forall z\exists w\neg R(x, y, z, w)$.

6.8 Equivalence of Tarskian and GTS truth (satisfaction)

The next theorem shows that Tarskian truth (falsity) is equivalent to game-theoretical truth (falsity):

Theorem 6.4 (Hintikka and Kulas, 1985, Hodges, 1989; Mann et al, 2011) *Let φ be a first-order formula, $\mathbb{M}$ a suitable model for the language of φ and s an assignment in M whose domain includes the free variables of φ. Then:*

1. $M, s \vDash^{+}_{GTS} \neg\varphi$ *iff* $M, s \nvDash^{+}_{GTS} \varphi$
2. $M, s \vDash^{+}_{GTS} (\varphi \wedge \psi)$ *iff* $M, s \vDash^{+}_{GTS} \varphi$ *and* $M, s \vDash^{+}_{GTS} \psi$
3. $M, s \vDash^{+}_{GTS} (\varphi \vee \psi)$ *iff* $M, s \vDash^{+}_{GTS} \varphi$ *or* $M, s \vDash^{+}_{GTS} \psi$
4. $M, s \vDash^{+}_{GTS} \exists x\varphi$ *iff* $M, s(x/a) \vDash^{+}_{GTS} \varphi$ *for some* $a \in M$
5. $M, s \vDash^{+}_{GTS} \forall x\varphi$ *iff* $M, s(x/a) \vDash^{+}_{GTS} \varphi$ *for every* $a \in M$

Proof We have already proved (1). For (2), assume that σ is a winning strategy for Eloise in $G(\mathbb{M}, s, \varphi \wedge \psi)$. Define winning strategies σ' and σ'' for Eloise in $G(\mathbb{M}, s, \varphi)$ and $G(\mathbb{M}, s, \psi)$, respectively, the following way:

$$\begin{aligned} \sigma'(s, \varphi, ...) &= \sigma(s, \varphi \wedge \psi, \varphi, ...) \\ \sigma''(s, \psi, ...) &= \sigma(s, \varphi \wedge \psi, \psi, ...) \end{aligned}$$

That is, σ' is the part of σ that Eloise consults when Abelard chooses φ and σ'' is the part of σ that she consults after Abelard chooses ψ. Obviously σ' is a winning strategy for $G(\mathbb{M}, s, \varphi)$, and σ'' is a winning strategy for $G(\mathbb{M}, s, \psi)$. Conversely, suppose Eloise has a winning strategy σ' in $G(\mathbb{M}, s, \varphi)$, and a winning strategy σ'' in $G(\mathbb{M}, s, \psi)$. We define a winning strategy σ for Eloise in $G(\mathbb{M}, s, \varphi \wedge \psi)$ as follows:

$$\begin{aligned} \sigma(s, \varphi \wedge \psi, \varphi, ...) &= \sigma'(s, \varphi, ...) \\ \sigma(s, \varphi \wedge \psi, \psi, ...) &= \sigma''(s, \psi, ...) \end{aligned}$$

That is, σ is the winning strategy for $G(\mathbb{M}, s, \varphi \wedge \psi)$ that tells Eloise to mimic σ' if Abelard chooses φ and to mimic σ'' if he chooses ψ.

For (3), suppose σ is a winning strategy for Eloise in $G(\mathbb{M}, s, \varphi \vee \psi)$. If $\sigma(s, \varphi \vee \psi) = \varphi$ then we define a strategy σ' for Eloise in the game $G(\mathbb{M}, s, \varphi)$ by:

$$\sigma'(s, \varphi, ...) = \sigma(s, \varphi \vee \psi, \varphi, ...)$$

That is, σ' is the strategy for $G(\mathbb{M}, s, \varphi)$ that mimics the part of σ that Eloise consults after choosing φ. Every play $h' = (s, \varphi, ...)$ corresponds to a play $h = (s, \varphi \vee \psi, \varphi, ...)$. Obviously σ' is winning given that σ is. There is a symmetric argument for the case in which $\sigma(s, \varphi \vee \psi) = \psi$. For the converse, assume either Eloise has a winning strategy σ' in $G(\mathbb{M}, s, \varphi)$, or a winning strategy σ'' in $G(\mathbb{M}, s, \psi)$. If the former, define a strategy σ for Eloise in $G(\mathbb{M}, s, \varphi \vee \psi)$ by:

$$\begin{aligned} \sigma(s, \varphi \vee \psi) &= \varphi \\ \sigma(s, \varphi \vee \psi, \varphi, ...) &= \sigma'(s, \varphi, ...) \end{aligned}$$

That is, σ tells Eloise to choose the disjunct for which she has a winning strategy (σ'), then follow that strategy. The case for the winning strategy σ'' in $G(\mathbb{M}, s, \psi)$ is similar.

(5) Suppose σ is a winning strategy for Eloise in $G(\mathbb{M}, s, \forall x\varphi)$. For every a we define a strategy σ_a for Eloise in the game $G(\mathbb{M}, s(x/a), \varphi)$:

$$\sigma_a(s(x/a), \varphi, ...) \quad = \quad \sigma(s, \forall x\varphi, (x, a), ...)$$

That is, σ_a tells Eloise to mimic the part of σ she consults when Abelard chooses a. Observe that σ_a is winning for $G(\mathbb{M}, s(x/a), \varphi)$. For the converse, suppose that for each a Eloise has a winning strategy σ_a in the game $G(\mathbb{M}, s(x/a), \varphi)$. We define a strategy σ for Eloise in $G(\mathbb{M}, s, \forall x\varphi)$ by:

$$\sigma(s, \forall x\varphi, (x, b), ...) \quad = \quad \sigma_b(s(x/b), \varphi, ...)$$

That is, σ tells Eloise to mimic σ_b when Abelard sets the value of x to be b. □

One may wonder though what is the difference between the Tarksi-type interpretation and the GTS for first-order languages. Here is one possible answer:

> The game-theoretical truth-definition serves the same purpose as the other (recursive) clauses in a usual (Tarski-type) truth-definition. The main difference is that, whereas the recursive clauses in a usual Tarski-type truth-definition work their way from the inside out, starting from the simplest formulas, a game-theoretical truth-definition works its way from the outside in, defining in effect a semantical analysis of the sentence in question. This makes a difference, for it means that GTS is independent of the assumption of compositionality (semantical context-independence), and works even when compositional fails...(Hintikka and Kulas, 1985, p. 4)

6.9 Semantic games for FOL: strategic form

Any semantic game $G(\mathbb{M}, s, \varphi)$, where φ is a first-order formula may be transformed into a strategic game $\Gamma(\mathbb{M}, s, \varphi)$ which has the form $\Gamma = (S_\exists, S_\forall, \mu_\exists, \mu_\forall)$ where:

- $S_\exists$ is the set of strategies of Eloise in the semantic game $G(\mathbb{M}, s, \varphi)$
- $S_\forall$ is the set of strategies of Abelard in the semantic game $G(\mathbb{M}, s, \varphi)$

When Eloise follows the strategy σ and Abelard follows the strategy τ in the semantic game $G(\mathbb{M}, s, \varphi)$ the result is a play (maximal history) in the semantic game $G(\mathbb{M}, s, \varphi)$. We denote this history by $\sigma \circ \tau$. It is either a win for Eloise or a win for Abelard in the game $G(\mathbb{M}, s, \varphi)$.

- $u_\exists$, the utility function for Eloise is defined in the following way. Let $\sigma \in S_\exists$ and $\tau \in S_\forall$. Then

$$\mu_\exists(\sigma, \tau) = u_\exists(\sigma \circ \tau)$$

that is, the payoff of Eloise when she plays the strategy σ and Abelard plays the strategy τ in the strategic game is the payoff the players receive for the play $\sigma \circ \tau$ in the semantic game.

- Analogously,
$$\mu_{\forall}(\sigma, \tau) = u_{\forall}(\sigma \circ \tau) = 1 - u_{\exists}(\sigma, \tau)$$
.

Example We consider the game $G(\mathbb{M}, \varnothing, \varphi)$ where φ is the sentence $\forall x \exists y x = y$ and $\mathbb{M} = \{a, b\}$. We have seen that there are four strategies for Eloise

$$\begin{array}{lllllll} \sigma_1(h_a) & = & (y,a) & \quad & \sigma_1(h_b) & = & (y,b) \\ \sigma_2(h_a) & = & (y,b) & & \sigma_2(h_b) & = & (y,a) \\ \sigma_3(h_a) & = & (y,a) & & \sigma_3(h_b) & = & (y,a) \\ \sigma_4(h_a) & = & (y,b) & & \sigma_4(h_b) & = & (y,b) \end{array}$$

that we abbreviate as aa, ba, aa, and bb, respectively. There are two strategies for Abelard, τ_a and τ_b,

$$\tau_a(\varphi, \varnothing) \quad = \quad (x, a) \quad \text{and} \quad \tau_a(\varphi, \varnothing) \quad = \quad (x, b)$$

that we abbreviate as a and b, respectively. The strategic game $\Gamma(\mathbb{M}, \varnothing, \varphi)$ may be presented as the matrix:

	a	b
ab	$(1,0)$	$(1,0)$
ba	$(0,1)$	$(0,1)$
aa	$(1,0)$	$(0,1)$
bb	$(0,1)$	$(1,0)$

To rehearse how the payoffs are computed, consider the strategy profile consisting of ab and a. When each player follows his strategy in the profile, the result $ab \circ a$ is the play
$$h_{aa} = (\varnothing, \varphi, (x, a), (y, a))$$
in the semantical game $G(\mathbb{M}, \varnothing, \varphi)$. Hence

$$\mu(ab, a) = u_{\exists}(ab \circ a) = u_{\exists}(h_{aa}) = 1.$$

We should be persuaded that for every strategy σ_i Eloise follows σ_i in $\sigma_i \circ \tau_1$ and $\sigma_i \circ \tau_2$. That is, Eloise follows σ_i in the play $\sigma_i \circ \tau$ of the semantic game $G(\mathbb{M}, \varnothing, \varphi)$, for every $\tau \in S_{\forall}$. And for every strategy τ_i Abelard follows τ_i in the play $\sigma \circ \tau_i$ for every $\sigma \in S_{\exists}$. Given that $\sigma_1 (= ab)$ is a winning strategy in $G(\mathbb{M}, \varnothing, \varphi)$, and that $\mu_{\exists}(\sigma_1, \tau) = u_{\exists}(\sigma_1 \circ \tau)$, it should come as no surprise that $\mu_{\exists}(\sigma_1, \tau) - 1$ for every strategy τ of Abelard. It also follows that every pair (σ_1, τ) is an equilibrium in the strategic game $\Gamma(\mathbb{M}, \varnothing, \varphi)$. This is easy to see: given that $\mu_{\exists}(\sigma_1, \tau) = 1$, then for every other strategy σ Eloise might play, she cannot get a bigger payoff. And given that $\mu_\forall(\sigma_1, \tau) = 0$, for every strategy τ of Abelard, Abelard cannot get a better payoff by unilateral deviation.

The above example should convince us of the truth of the following proposition:

Proposition 6.5 Let φ be a first-order formula, $\mathbb{M}$ a model and s an assignment whose domain contains the free variables of φ. Let $G(\mathbb{M}, \varnothing, \varphi)$ denote the corresponding extensive game and $\Gamma(\mathbb{M}, \varnothing, \varphi)$ denote the corresponding strategic game. Then:

(i) Eloise has a winning strategy in $G(\mathbb{M}, \varnothing, \varphi)$ if and only if $\Gamma(\mathbb{M}, \varnothing, \varphi)$ has an equilibrium which returns to Eloise the payoff 1 and to Abelard the payoff 0.

(ii) Abelard has a winning strategy in $G(\mathbb{M}, \varnothing, \varphi)$ if and only if $\Gamma(\mathbb{M}, \varnothing, \varphi)$ has an equilibrium which returns to Abelard the payoff 1 and to Eloise the payoff 0.

Chapter 7

Independence-friendly logic

7.1 Introduction

Let us return to our earlier game associated with the first-order sentence

$$\forall x \exists y \forall x' \exists y' Q(x, x', y, y')$$

Suppose that instead of having y' depending on both x and x', we want y' to depend only on x' (and keep the other dependencies intact). Game-theoretically this will correspond to the idea that in her strategies, Eloise will not be allowed to use information about the choice for x but only information about the choice for x' and information about her own choice for y. To re-establish the correspondence between syntax and the information sets of the players in the underlying game, Hintikka and Sandu (1989) introduced a new notation

$$\forall x \exists y \forall x' (\exists y' / \{x, y\}) Q(x, x', y, y')$$

Intuitively: when choosing a value for y' Eloise does not know the choice of her opponent for x neither her own earlier choice for y. The fact that the players ignore some of their earlier choices is known in the literature as imperfect recall. It makes the playability of the games rather hard. An alternative interpretation is given by Barwise (in the context of branching quantifiers). We think of the players as forming teams, each occurrence of the universal quantifier or conjunction being one player in Abelard's team, and each occurrence of the existential quantifier or disjunction being a player in Eloise's team. The interaction of the players in the same team gives rise to the phenomenon of signaling. Here is one example (Janssen and Duchesne). We show that the IF sentence

$$\forall x_0 \forall x_2 (x_0 \neq x_2 \vee (\exists x_1 / \{x_0\}) x_0 = x_1)$$

is a logical truth by showing that the team of Eloise, consisting of two players, has a winning strategy on every model (which has at least two individuals):

- Let Abelard choose a_0 and a_2.
- If $a_0 \neq a_2$, then the first player in Eloise team chooses Left. Eloise's team wins.
- If $a_0 = a_2$, then the same player chooses Right and then the second player chooses a_2.

Notice that despite the second player in the team of Eloise not knowing the choice of the first player in Abelard's team, the first player knows it, and signals it to her team mate: choosing Right is a way for her to say "choose the same as the second player in Abelard's team" (the second player in Eloise's team knows that choice). We now introduce the syntax and the semantics of IF logic in a precise way.

7.2 The syntax

The independence-friendly formulas, IF formulas, for short, are formed in the following way (we assume an underlying vocabulary L):

- $t_1 = t_2$ and $\neg t_1 = t_2$ ($t_1 \neq t_2$) are IF formulas
- If R is an n-ary relation symbol and $t_1, ..., t_n$ are terms, then $R(t_1, ..., t_n)$ and $\neg R(t_1, ..., t_n)$ are IF formulas
- If φ and ψ are IF formulas, so are $(\varphi \vee \psi)$ and $(\varphi \wedge \psi)$.
- If φ is an IF formula and W is a finite set of variables, then $(\exists x/W)\varphi$ and $(\forall x/W)\varphi$ are IF formulas .

The interpretation of e.g. $(\exists x/W)$ is: "x is independent of the values of the variables in W". When $W = \varnothing$, we recover the standard quantifiers and connectives. Notice that we restrict attention to IF formulas in negation normal form. In this chapter we shall consider only standard disjunctions and conjunctions.

The set of free variables of an IF formula φ is defined as in ordinary first-order logic, except for the clauses for the quantifiers which are now replaced by:

$$Free((\exists x/W)\varphi) \quad = \quad Free(\forall x/W)\varphi) \quad = \quad (Free(\varphi) - \{x\}) \cup W.$$

Thus in the formula $(\exists x_1/\{x_1, x_2\})\, R(x_1, x_2)$ the variable x_2 is free.

An occurrence of a variable x is bound by the innermost quantifier (Qx/W) in the scope of which it occurs. For instance in the formula

$$\forall x\, (\exists y/\{x\})\, R(x, y) \wedge \forall y\, (\exists z/\{x, y\})\, R(y, z)$$

the variable x is both free and bound while the variables y and z are bound.

7.3 Semantical games: extensive games of imperfect information

The notions of models, assignments and the satisfaction clauses for atomic formulas are the same as for ordinary first-order languages. The game-theoretical interpretation of the latter naturally extends to the present case: the assumption of the players' information in finite 2-player, win-lose extensive games is now relaxed to imperfect information.

The rules of a semantical game $G(\mathbb{M}, s, \varphi)$ where φ is an IF formula in negation normal form, $\mathbb{M}$ is a model which interprets the vocabulary of φ, and s is an assignment in $\mathbb{M}$ which includes the free variables of φ are identical with the five rules given earlier: in the last two clauses $(\exists x\psi, r)$ and $(\forall x\psi, r)$ are replaced with $((\exists x/W)\psi, r)$ and $((\forall x/W)\psi, r)$, respectively. The changes will affect only the information sets and thereby the strategies of the players. The imperfect information comes in the form of the players' restricted access to the current assignment in the game.

Let W be a set of finite variables, s and s' be two assignments in a model $\mathbb{M}$ with the same domain which includes W. We say that that s and s' are W-equivalent, $s \approx_W s'$, if for every variable $x \in dom(s) - W$ we have $s(x) = s'(x)$.

Example 7.1 Let s and s' two assignments such that

$$dom(s) = dom(s') = \{x_0, ..., x_7\}$$

and $s(x_i) = a_i$ and $s'(x_i) = b_i$. Let $W = \{x_2, x_5\}$. The assignments s and s' are are $W-$equivalent, exactly when

$$a_i = b_i$$

for every $i = 0, 1, 3, 4, 6, 7$.

Notice that when $W = dom(s) = dom(s')$, then the above condition is trivially satisfied, for $dom(s) - W = \varnothing$.

Every history h in $G(\mathbb{M}, s, \varphi)$ induces an assignment s_h in the model $\mathbb{M}$, which extends or modifies the initial assignment s. We are going to define two indistinguishable relations $\sim_\exists$ and $\sim_\forall$ on the set of non-maximal histories of the game.

For $\sim_\exists$ there are two cases:

- It is Eloise's turn to move corresponding to a disjunction $\psi \vee \psi'$. In the context of the extensive game, let h and h' be two histories in the game where such a move is supposed to happen. Then we stipulate that

$$h \sim_\exists h' \quad \Leftrightarrow \quad s_h = s_{h'}.$$

- It is Eloise's turn to move corresponding to $(\exists x/W)\psi$. Then for any two histories h, h' where such a move is about to happen, we stipulate:

$$h \sim_\exists h' \quad \Leftrightarrow \quad s_h \approx_W s_{h'}.$$

For the relation $\sim_\forall$ we also have two cases:

- It is Abelard's turn to move corresponding to a conjunction $\psi \wedge \psi'$. In the context of the extensive game, let h and h' be two histories in the game where such a move is supposed to happen. Then we stipulate that

$$h \sim_\wedge h' \quad \Leftrightarrow \quad s_h = s_{h'}.$$

- It is Abelard's turn to move corresponding to $(\forall x/W)\psi$. Then for any two histories h, h' where such a move is about to happen, we stipulate:

$$h \sim_\forall h' \quad \Leftrightarrow \quad s_h \approx_W s_{h'}.$$

The relations $\sim_\exists$ and $\sim_\forall$ specify exactly how much information the players have at their disposal at a given decision point.

The notions of winning and loosing a play (maximal history) of the game are defined exactly as in the case of games of perfect information.

A strategy σ_p for player p in the semantical game $G(\mathbb{M}, s, \varphi)$ is defined exactly as in the games of perfect information, except for the requirement of uniformity: for every h, h'

$$h \sim_p h' \quad \Rightarrow \quad \sigma_p(h) = \sigma_p(h').$$

The notion of truth and falsity are defined as before:

Definition 7.1 Let φ be an IF formula, $\mathbb{M}$ a model and s an assignment in M whose domain contains $Free(\varphi)$. Then

- $\mathbb{M}, s \models^+_{GTS} \varphi$ iff there is a winning strategy for Eloise in $G(\mathbb{M}, s, \varphi)$
- $\mathbb{M}, s \models^-_{GTS} \varphi$ iff there is a winning strategy for Abelard in $G(\mathbb{M}, s, \varphi)$.
- An IF sentence φ is true in M, $\mathbb{M} \models^+_{GTS} \varphi$ if and only if $\mathbb{M}, \varnothing \models^+_{GTS} \varphi$
- An IF sentence φ is false in M, $\mathbb{M} \models^-_{GTS} \varphi$ if and only if $\mathbb{M}, \varnothing \models^-_{GTS} \varphi$.

7.4 Indeterminacy and Signaling in IF logic

In games of imperfect information we cannot any longer rely on Zermelo's theorem to guarantee that one of the two players has a winning strategy in the game. Here we have an example of an indeterminate sentence which is nothing else than the Matching Pennies game we have analyzed earlier.

Example 7.2 (Matching Pennies) The Matching Pennies game that we have encountered several times in this can be expressed in IF logic using the IF sentence

$$\varphi_{MP}, \varphi_{MP} = \forall x(\exists y/\{x\})\, x = y$$

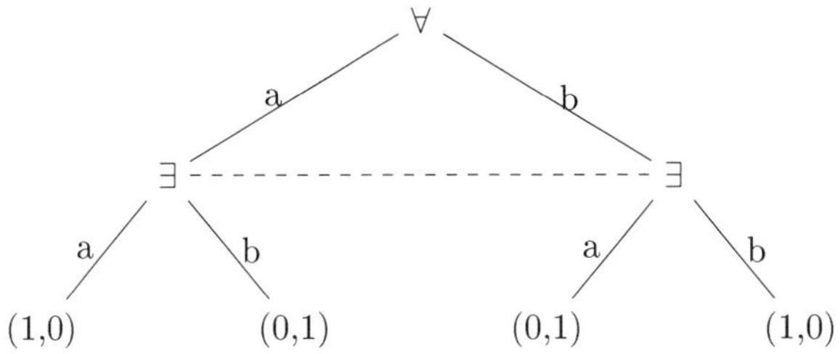

Figure 7.1: The game of Matching Pennies

interpreted in a two element model $\mathbb{M} = \{a, b\}$. The extensive game can be represented in a tree form as in figure 7.1.

where the dots indicate that the two histories

$$\begin{aligned} h_a &= (\varnothing, \varphi_{MP}, (x, a)) \\ h_b &= (\varnothing, \varphi_{MP}, (x, b)) \end{aligned}$$

are equivalent for Eloise, i.e., $h_a \sim_\exists h_b$. Note that their equivalence is due to the trivial $\{x\}$ –equivalence of the corresponding assignments $s_{h_a} = \{(x, a)\}$ and $s_{h_b} = \{(x, b)\}$. It follows from our earlier analysis of the game that neither Eloise nor Abelard has a winning strategy in the game:

$$\mathbb{M} \nvDash^+_{GTS} \varphi \quad \text{and} \quad \mathbb{M} \nvDash^-_{GTS} \varphi.$$

We now show an example where adding a dummy quantifier changes the truth-value of a formula. This illustrates the phenomenon of signaling in IF logic.

Example 7.3 (Hodges (1997)) This example discusses the following sentence φ_H (Hodges sentence)

$$\forall x \exists z (\exists y / \{x\}) x = y$$

which is formed from the Matching Pennies sentence φ_{MP} by inserting the dummy quantifier $\exists z$. Surprisingly Eloise has a winning strategy (on any model $\mathbb{M}$ with at least two elements) whereas she does not have one in the game $G(\mathbb{M}, \varnothing, \varphi_{MP})$, as shown earlier. Note that corresponding to $\exists z$, there are two histories where Eloise is to move

$$\begin{aligned} h_a &= (\varnothing, \varphi_H, (x, a)) \\ h_b &= (\varnothing, \varphi_H, (x, b)) \end{aligned}$$

and corresponding to $(\exists y / \{x\})$, there are four histories where she is to move:

$$\begin{aligned} h_{aa} &= (\varnothing, \varphi_H, (x, a), (z, a)) \\ h_{ba} &= (\varnothing, \varphi_H, (x, b), (z, a)) \\ h_{ab} &= (\varnothing, \varphi_H, (x, a), (z, b)) \\ h_{bb} &= (\varnothing, \varphi_H, (x, b), (z, b)). \end{aligned}$$

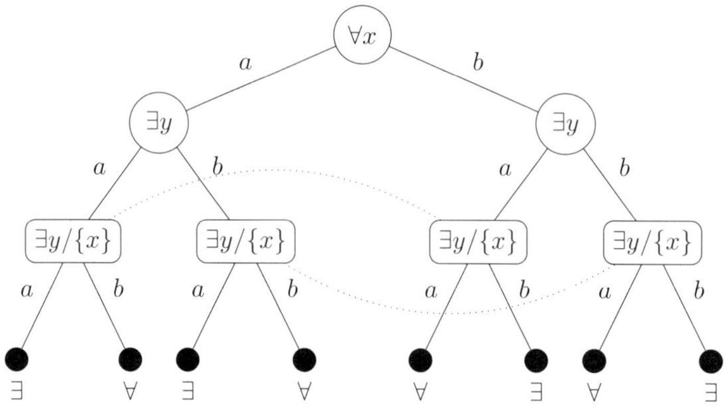

Figure 7.2: The semantic game for $\forall x \exists y(\exists y/\{x\})x = y$ in $M = \{a, b\}$

We can represent the game in tree form as in figure 7.2

We observe that that $s_{aa} \approx_{\{x\}} s_{ba}$, and $s_{ab} \approx_{\{x\}} s_{bb}$, hence $h_{aa} \sim_{\exists} h_{ba}$ and $h_{ab} \sim_{\exists} h_{bb}$. Therefore by the uniformity requirement, Eloise's strategies σ must be such that $\sigma(h_{aa}) = \sigma(h_{ba})$ and $\sigma(h_{ab}) = \sigma(h_{bb})$. Here is a winning strategy for Eloise:

$$\sigma(h_a) = (y, a) \qquad \sigma(h_{aa}) = \sigma(h_{ba}) = (y, a)$$
$$\sigma(h_b) = (y, b) \qquad \sigma(h_{ab}) = \sigma(h_{bb}) = (y, b)$$

There are two maximal histories in which Eloise follows this strategy

$$h_{aaa} = (\varnothing, \varphi_H, (x, a), (z, a), (y, a)) \quad \texttt{and} \quad h_{bbb} = (\varnothing, \varphi_H, (x, b), (z, b), (y, b))$$

and she wins both of them. The strategy of Eloise may be described as follows: wait to see the choice of Abelard, then repeat it, and finally repeat one's own choice. Notice that this makes sense, given that in order to repeat Abelard choice, Eloise has to see it, and she does, and in order to repeat her own choice, she has to see it (remember it), which she also does.

The above example illustrates the game-theoretical phenomenon of signaling: the players play according to a predetermined convention in order to convey information about their holdings. In our example Eloise does not "see" the choice of Abelard when making her second move. Some remarks are in order, at the point, about the knowledge or the memory of the players. How can Eloise "forget" the first choice of Abelard once she knew it for her first move? Computer scientists speak about games of imperfect recall (see below). For the moment let us look at an alternative interpretation in terms of teams of players. (Barwise?). Abelard's team consists of so many players as there are occurrences of universal quantifiers and conjunctions in the formula; analogously for Eloise's team. It is the team interpretation which generates the phenomenon of signaling. Signaling in IF logic has been extensively discussed (Caicedo et al, 2009; Janssen & Dechesne 2006).

Finally we give one more example of the phenomenon of signaling that now takes place among the players in Abelard's team (this is a variation of an example from Janssen & F. Dechesne, 2006).

Example 7.4 We will show that the sentence

$$\exists x \exists y \exists z[x = y \wedge \forall v(\forall u/\{x\})(u \neq x \vee v \neq z)]$$

is logically false, i.e. false on all models (recall we exclude the models with one element). Let $\mathbb{M}$ be an arbitrary model which contains at least two elements. We may think of $\wedge$, $\forall v$ and $\forall u$ as a team of universal players. Here is a winning strategy for this team.

- Let Eloise choose $a, b, c \in M$ as the values of x, y, and z, respectively
- If $a \neq b$ then $\wedge$ chooses Left; If $a = b$ then $\wedge$ chooses Right. After this $\forall v$ chooses c to be the value of v; finally $\forall u$ chooses b to be the value of u.

Notice that $\forall v$ "sees" the value of x; and $\forall u$ sees the value of y.

7.5 Negation in IF logic

Negation in IF logic is treated exactly in the same way as negation in ordinary first-order logic: it indicates roles swapping. The notion of strategy for a player p is defined exactly as before, as are game-theoretical truth and falsity.

Example 7.5 Consider the semantic game $G(\mathbb{M}, \varnothing, \neg\varphi_{MP})$ where φ_{MP} is the sentence $\forall x(\exists y/\{x\})\, x = y$ and $\mathbb{M} = \{a, b\}$. This example is identical to the earlier one ???, except for the independence of the existential quantifier. In the initial position $(\varnothing, \neg\varphi_{MP})$ Eloise is the verifier and Abelard is the falsifier. After the players switch roles, Eloise can choose either a or b which generates two histories:

$$\begin{aligned} h_a^* &= (\varnothing, \neg\varphi, \varphi, (x, a)) \\ h_b^* &= (\varnothing, \neg\varphi, \varphi, (x, b)) \end{aligned}$$

Each of them is a choice point for the verifier, Abelard, who can choose either a or b. This results in four maximal histories (plays)

$$\begin{aligned} h_{aa}^* &= (\varnothing, \neg\varphi, \varphi, (x, a), (y, a)) \\ h_{ab}^* &= (\varnothing, \neg\varphi, \varphi, (x, a), (y, b)) \\ h_{ba}^* &= (\varnothing, \neg\varphi, \varphi, (x, b), (y, a)) \\ h_{bb}^* &= (\varnothing, \neg\varphi, \varphi, (x, b), (y, b)). \end{aligned}$$

such that Abelard wins h_{aa}^* and h_{bb}^* and Eloise wins the remaining two. Given the uniformity requirement, Abelard has now two strategies (instead of four):

$$\begin{aligned} \tau_1(h_a^*) &= \tau_1(h_b^*) = (y, a) \\ \tau_2(h_a^*) &= \tau_1(h_a^*) = (y, b) \end{aligned}$$

None of them is winning. Abelard follows τ_1 it in the plays h^*_{aa} and h^*_{ba} but the second is a loss for her. And she follows τ_2 in the plays h^*_{ab} and h^*_{bb} but he looses the first. In a similar way one can show that there is no winning strategy for Eloise either.

Although we are still inside the class of 2-player, win-lose extensive games, the lack of perfect information prevents us, as we have seen, to apply Zermelo's theorem and prove

$$\mathbb{M}, s \models^+_{GTS} \neg\varphi \quad \texttt{iff} \quad \mathbb{M}, s \not\models^+ \varphi.$$

The direction from left-to-right

$$\mathbb{M}, s \models^+_{GTS} \neg\varphi \quad \Rightarrow \quad \mathbb{M}, s \not\models^+ \varphi$$

always hold, but the above example is a counter-example to the other direction. We know there is no winning strategy for Eloise in the semantic game $G(\mathbb{M}, s, \varphi_{MP})$. Our last example showed that there is no winning strategy for Eloise either in the game $G(\mathbb{M}, \varnothing, \neg\varphi_{MP})$.

We could introduce explicitly contradictory negation in IF logic by an explicit semantic clause:

$$\begin{array}{lll} \mathbb{M}, s \models^+_{GTS} \sim \varphi & \texttt{iff} & \mathbb{M}, s \not\models^+ \varphi \\ \mathbb{M}, s \models^-_{GTS} \sim \varphi & \texttt{iff} & \mathbb{M}, s \not\models^- \varphi. \end{array}$$

The last example showed that

$$\mathbb{M} \models^+_{GTS} \sim \varphi_{MP} \quad \texttt{and} \quad \mathbb{M} \not\models^+_{GTS} \neg\varphi_{MP}$$

that is, game theoretical negation and contradictory negation are not co-extensional in (extended) IF logic. One thing which needs to be emphasized is that contradictory negation $\sim$, unlike game theoretical negation $\neg$ can appear only sentence initially: the latter is interpreted by a game rule, while the former is interpreted by a meta-rule which refers to the (non)existence of certain strategies in the corresponding game. We shall not pursue the matter of contradictory negation any longer here.

Let us return to IF logic and game-theoretical negation. Although the principle of determinacy fails, the other properties of game-theoretical negation still hold. That is, for every IF formula φ, model $\mathbb{M}$ and assignment s in $\mathbb{M}$ whose domain includes the free variables of φ we have:

(i) There is a winning strategy for Eloise in $G(\mathbb{M}, s, \neg\varphi)$ if and only if there is a winning strategy of Abelard in $G(\mathbb{M}, s, \varphi)$:

$$\mathbb{M}, s \models^+_{GTS} \neg\varphi \quad \texttt{iff} \quad \mathbb{M}, s \models^-_{GTS} \varphi$$

(ii) There is a winning strategy for Eloise in $G(\mathbb{M}, s, \varphi)$ if and only if there is a winning strategy of Abelard in $G(\mathbb{M}, s, \neg\varphi)$:

$$\mathbb{M}, s \models^+_{GTS} \varphi \quad \texttt{iff} \quad \mathbb{M}, s \models^-_{GTS} \neg\varphi.$$

One can also show that the double negation law, de Morgan laws and the lwas governing the the interchange of quantifiers and negation hold in IF logic too:

$$\begin{array}{rcl}
\mathbb{M}, s \models^{\pm}_{GTS} \neg\neg\varphi & \texttt{iff} & \mathbb{M}, s \models^{\pm}_{GTS} \varphi \\
\mathbb{M}, s \models^{\pm}_{GTS} \neg(\varphi \wedge \psi) & \texttt{iff} & \mathbb{M}, s \models^{\pm}_{GTS} (\neg\varphi \vee \neg\psi) \\
\mathbb{M}, s \models^{\pm}_{GTS} \neg(\varphi \vee \psi) & \texttt{iff} & \mathbb{M}, s \models^{\pm}_{GTS} (\neg\varphi \wedge \neg\psi) \\
\mathbb{M}, s \models^{\pm}_{GTS} \neg(\exists x/W)\varphi & \texttt{iff} & \mathbb{M}, s \models^{\pm}_{GTS} (\forall x/W)\neg\varphi \\
\mathbb{M}, s \models^{\pm}_{GTS} \neg(\forall x/W)\varphi & \texttt{iff} & \mathbb{M}, s \models^{\pm}_{GTS} (\exists x/W)\neg\varphi.
\end{array}$$

By a repeated application of these equivalences one can show that any IF formula is both true and falsity equivalent to an IF formula in negation normal form, i.e. a formula in which the negation sign is only prefixed to atomic formulas.

7.6 IF prefixes which are not first-order

The basic motivation for introducing IF logic is the limited expressive power of ordinary first-order logic, that is, its incapacity to express some basic mathematical concepts. Here is an example which shares the same worry:

> ...I recently came across the phenomenon of nonfirstorderisability in mathematical logic: there are perfectly meaningful and useful statements in mathematics which cannot be phrased within the confines of first order logic (combined with the language of set theory, or any other standard mathematical theory); one must use a more powerful language such as second order logic instead. This phenomenon is very well known among logicians, but I hadn't learned about it until very recently, and had naively assumed that first order logic sufficed for "everyday" usage of mathematics (Terrence Tao, 2007)

Tao notices that one can express in first-order logic statements like:

- For every x, there exists a y depending on x such that $B(x, y)$ is true

by

$$\forall x \exists y B(x, y)$$

(here $B(x, y)$ is a binary relation.)

He further observes that we can also express the statement

- For every x, there exists a y independent of x such that $B(x, y)$ is true

by

$$\exists y \forall x B(x, y).$$

One can express even more complex statements involving four quantifiers like

- For every x and x', there exists a y depending only on x and a y' depending on x and x' such that $Q(x, x', y, y')$ is true by

$$\forall x \exists y \forall x' \exists y' Q(x, x', y, y')$$

(here $Q(x, x', y, y')$ is a quaternary relation;note that this allows y' to depend on y also, but this turns out to be moot, because y depends only on x)

But Tao notices that one cannot always express

- For every x and x', there exists a y depending only on x and a y' depending only on x' such that $Q(x, x', y, y')$ is true

  His conclusion is:

  It seems to me that first order logic is limited by the linear (and thus totally ordered) nature of its sentences; every new variable that is introduced must be allowed to depend on all the previous variables introduced to the left of that variable. This does not fully capture all of the dependency trees of variables which one deals with in mathematics. (Terrence Tao).

The example given by Tao illustrates one of the two kinds of configurations of quantifiers which cannot be always expressed in ordinary first-order logic.

We say that a quantifier (Qx/X) *depends* on (Qy/Y)

$$...(Qy/Y)...(Qx/X)...$$

whenever (Qx/X) *is in the scope of* (Qy/Y) and in addition,$y \notin X$. If (Qx/X) does not depend on (Qy/Y), then we say that it is *independent* of (Qy/Y).

So notice that (Qx/X) is independent of (Qy/Y) either when (Qx/X) is not in the scope of (Qy/Y), or, it is in the scope of (Qy/Y) but $y \in X$.

An IF prefix is a *Henkin prefix* if it contains 4 quantifiers

$$...\forall x...\forall y...(\exists u/U)...(\exists v/V)...$$

in any order such that:

- $(\exists u/U)$ depends on $\forall x$ and is independent of $\forall y$ and $(\exists v/V)$
- $(\exists v/V)$ depends on $\forall y$ and is independent of $\forall x$ and $(\exists u/U)$.

The idea here is that each of the existential quantifiers depends only on one of the two universal quantifiers (hence it is independent of the other existential quantifier).

1. $\forall x \exists u \forall y (\exists v/\{x, u\})$
2. $\forall x \forall y (\exists u/\{y\})(\exists v/\{x, u\})$
3. $\forall x \exists u (\forall y/\{x\})(\exists v/\{x, u\})$
4. $\forall x \exists u (\forall y/\{u\})(\exists v/\{x, u\})$

5. $\forall x \exists u (\forall y / \{x, u\})(\exists v / \{x, u\})$

6. $\forall x (\forall y / \{x\})(\exists u / \{y\})(\exists v / \{x, u\})$

Actually it can be shown every sentential Henkin prefix extends, modulo the renaming of bound variables, one of the prefixes (1)-(6), where an extension of a quantifier prefix is any embedding of the latter into a larger prefix which preserves its dependence relations. For instance,

7. $\exists t \forall x (\exists u / \{t\}) \forall y (\exists v / \{x, u, \})$

8. $\forall z \forall x \exists u \forall t \forall y \exists w (\exists v / \{x, u\})$

are extensions of (1).

Note that the configuration invoked by Tao is an example of (1) or (2). (Actually all the prefixes 1- 6 are truth-equivalent).

We shall look at an example of a Henkin prefix having greater expressive power than ordinary first-order logic after we deal with the next case. It turns out that one can exceed the expressive power of first-order logic with a configuration of three quantifiers.

An IF prefix is a *signaling prefix* if it contains three quantifiers

$$...(\forall u / U)...(\exists v / V)...(\exists w / W)...$$

such that:

- $(\exists v / V)$ depends on $(\forall u / U)$
- $(\exists w / W)$ depends on $(\exists v / V)$ but is independent of $(\forall u / U)$

The idea in a signaling prefix is that the second existential quantifier does not "see" the universal quantifier, but it sees another existential quantifier which, in turn, sees the universal quantifier. Notice that whereas in Henkin prefixes the existential quantifiers are independent of each other, in signaling prefixes, one of the existential quantifier is dependent on the other. We have seen an example of a signaling sequence when discussing Hodges' example (here example 7 below). Actually it can be shown that modulo the renaming of bound variables, every sentential signaling prefix is an extension of the Hodges' sequence (9). Here are few more examples:

9. $\forall u \exists v (\exists w / \{u\})$

10. $\forall u \forall z \exists v (\exists w / \{u\})$

11. $\forall u \forall z \exists v (\exists w / \{u, z\})$

12. $\forall y \exists u \forall x \exists w (\exists v / \{y\})$

Note that in the last example $\exists u$ depends on $\forall y$ and $(\exists v / \{y\})$ depends on $\exists u$ but is independent of $\forall y$.

We call an IF prefix *primary* if it is neither Henkin nor signaling.

Both Henkin and signaling prefixes leads to to expressive power that is greater than that of ordinary, first-order logic. Here is our main example.

7.7 Expressing infinity

We say that a set M is infinite if there is a function $h : M \rightarrow M$ such that:

1. h is injective: For all $a, b \in M$, $a \neq b \Rightarrow h(a) \neq h(b)$; and
2. There is $c \in M$ such that for all $a \in M : f(a) \neq c$.

We can visualize the situation in the following way. Think of M as the set of natural numbers, $M = \{0, 1, 2, ...\}$. Here is an example of a function h which fulfills conditions (1) and (2)

$$h(0) = 1, \quad h(1) = 2, \quad h(2) = 3, \quad \ldots$$

In this case $c = 0$.

The IF sentence that defines infinity is φ_∞

$$\forall x \exists y (\exists z / \{x\})(y \neq c \wedge z = x).$$

We show that

(*) $M \models^+ \varphi_\infty$ if and only if there is a function $h : M \rightarrow M$ such that conditions (1) and (2) are satisfied.

Assume that $\mathbb{M} \models^+ \varphi_\infty$. By definition, there is a winning strategy σ for Eloise in $G(M, \varnothing, \varphi_\infty)$. That is, there is a function σ such that for every $a \in M$

$$\sigma(\varphi_\infty, \varnothing, (x, a))$$

is a value for y and for every $b \in M$

$$\sigma(\varphi_\infty, \varnothing, (x, a), (y, b))$$

is a value for z. Notice that for every $a \in M$ there is exactly one b such that

$$\sigma(\varphi_\infty, \varnothing, (x, a)) = (y, b)$$

and for every $a, d \in M$ there is exactly one $e \in M$ such that

$$\sigma(\varphi_\infty, \varnothing, (x, a), (y, d)) = (z, e).$$

Both claims follow from the fact that the strategy σ is a function. The prefix $\forall x \exists y (\exists z / \{x\})$ is identical to the prefix $\forall x \exists y \exists z$ in our earlier example ???. In the same way as we did in that example, we can associate with σ two functions $f : M \rightarrow M$ and $g : M^2 \rightarrow M$. We let f be the function corresponding to $\exists y$ such that for every $a \in M$

$$\sigma(\varphi, \varnothing, (x, a)) = (y, f(a)).$$

We let g be the function corresponding to $\exists z$ be such that for every $a, b' \in M$ we have

$$\sigma(\varphi, \varnothing, (x, a), (y, b)) = (z, g(a, b)).$$

There is however a basic difference with the earlier example. The independence $(\exists z/\{x\})$ imposes a condition of uniformity on the strategy σ and consequently on the function g. Consider any two histories

$$\begin{aligned} h &= (\varphi, \varnothing, (x,a), (y,b)) \\ h' &= (\varphi, \varnothing, (x,a'), (y,b)) \end{aligned}$$

and the two assignments they induce

$$\begin{aligned} s_h &= \{(x,a), (y,b)\} \\ s_{h'} &= \{(x,a'), (y,b)\} \end{aligned}$$

s_h and $s_{h'}$ are $\{x\}$-equivalent, for they agree on the values of the variables other than x (i.e. they agree on y). But then h and h' are equivalent and we must have

$$\sigma(h) = \sigma(h').$$

But then, by the Uniformity condition, we must also have:

$$g(a,b) = g(a',b)$$

for every $a, a' \in M$. In other words, the values of g depend only on b and we can take g to be a function only of one argument, b.

Eloise follows the strategy σ, or alternatively, the pair of functions (f,g), in all the plays of the form:

$$(\varphi, \varnothing, (x,a), (y, f(a)), (z, g(f(a)))).$$

If, in addition, the strategy σ is winning, we must have

$$g(f(a)) = a \quad \texttt{and} \quad f(a) \neq c^M,$$

for every $a \in M$. In other words, $f : M \to M$ is an injection whose range is distinct from c^M.

For the converse, suppose M is infinite. Then there is a function $f : M \to M$ which is an injection and such that for all $a \in M$, $f(a) \neq c^M$. We define a strategy σ for Eloise in the game $G(M, \varnothing, \varphi_\infty)$ in the following way:

$$\sigma(\varphi_\infty, \varnothing, (x,a)) = (y, f(a))$$

and for any $b \in M$

$$\sigma(\varphi_\infty, \varnothing, (x,a), (y,b)) = f^{-1}(b)$$

if b is in the range of f; and c^M otherwise.

We next show that σ is a winning strategy. Let

$$h_a = (\varphi_\infty, \varnothing, (x,a), (y,b), (z,d))$$

where

$$(y,b) = \sigma(\varphi_\infty, \varnothing, (x,a))$$

and

$$(z,d) = \sigma(\varphi_\infty, \varnothing, (x,a), (y,b))$$

be a play of the game where σ is used. We know that $b = f(a) \neq c^M$. It remains to show that $\sigma(\varphi_\infty, \varnothing, (x,a), (y,b)) = (z,d)$ where $b = f(a)$ is such that $d = a$. By the definition of σ :

$$\sigma(\varphi_\infty, \varnothing, (x,a), (y,b)) = f^{-1}(b) = a$$

and we are done.

In the previous example we needed two functions f and g to simultate a function h which is an injection. We can obtain the same result with another IF sentence which states that there are two functions which denote the same injective function. Let φ_H denote the IF sentence

$$\exists t \forall x \forall y (\exists u / \{y,t\})(\exists v / \{x,u,t\}) \left[(x = y \leftrightarrow u = v) \wedge u \neq t\right].$$

As in the previous case, the strategy of Eloise can be decomposed into two function f_u and g_v, corresponding to the two existential quantifiers. Given the uniformity conditions, these functions are unary (syntactically f takes only x as an argument and g takes only y as an argument). The left-to-right implication in the first conjunct in the quantifier-free formula states that whenever $a = b$ then $f(a) = g(b)$, that is, f and g are the same function, say f. The right-to-left direction, reformulated as $x \neq y \rightarrow u \neq v$, states that f is an injection, and the last conjunct states that $f(a) \neq t$.

Notice that the prefix of the sentence φ_H is a Henkin prefix.

7.8 The incompleteness of IF logic

It follows, from the result in the previous section that IF logic is incomplete. We rehearse the argument.

Let φ_n be the FOL sentence which asserts that the universe has at least n elements:

$$\exists x_1 ... \exists x_n \left(\bigwedge_{1 \leq i \leq n} x_i \neq x_j \right).$$

It is easy to see that

$$\{\varphi_n : n \geq 2\} \models^+ \varphi_\infty.$$

However, there is no finite $\Delta \subseteq \{\varphi_n : n \geq 2\}$ such that $\triangle \models^+ \varphi_\infty$.

On the basis of this example one can show that IF logic is semantically incomplete:

Theorem 7.2 *There is no sound and semantically complete proof system for IF logic. In other words, there is no proof system $\vdash_{IF}$ such that for every set of IF sentences Γ and IF sentence φ we have*

$$\Gamma \vdash_{IF} \varphi \quad iff \quad \Gamma \models^{+} \varphi.$$

PROOF Suppose that there is such a system and take Γ to be the set $\{\varphi_n : n \geq 2\}$ and φ to be our sentence φ_∞. Given that $\{\varphi_n : n \geq 2\} \models^{+} \varphi_\infty$ we must also have $\{\varphi_n : n \geq 2\} \vdash_{IF} \varphi_\infty$. But the notion $\vdash_{IF}$ is finitistic, i.e. $\{\varphi_n : n \geq 2\} \vdash_{IF} \varphi_\infty$ must use only a finite number of premises $\Delta \subseteq \{\varphi_n : n \geq 2\}$. Thus $\Delta \vdash_{IF} \varphi_\infty$, which by the above equivalence implies $\Delta \models^{+} \varphi_\infty$, a contradiction. □

7.9 Equivalent IF prefixes

Let φ and φ' be two IF formulas. We say that φ and φ' are equivalent if for every model $\mathbb{M}$ and assignment s which includes $free(\varphi)$ and $free(\varphi')$ we have:

$$\mathbb{M}, s \models^{+}_{GTS} \varphi \quad \text{iff} \quad \mathbb{M}, s \models^{+}_{GTS} \varphi'$$

Two prefixes Π and Π' of IF quantifiers are said to be *equivalent* if $\Pi\varphi$ and $\Pi'\varphi$ are equivalent for every quantifier-free formula φ.

Recall that throughout the book we do not allow for double quantification:

- No quantifier of the form (Qx/W) is in the scope of a quantifier of the form $(\exists x/V)$ or $(\forall x/V)$.

A prefix $\Pi = (Q_1u_1/U_1)...(Q_nu_n/U_n)$ of IF quantifiers is *sentential* if $U_i \subseteq u_1, ..., u_{i-1}$ for every $1 \leq i \leq n$. So notice that if a prefix is sentential, then we must have $U_1 = \varnothing$ and then, for each quantifier (Q_iu_i/U_i), $i > 1$, all the variables in the slash set U_i must be bound by quantifiers which occur earlier in the sequence.

Drawing on previous work in the IF literature, Sevenster (2014) identifies 3 operations on prefixes of IF quantifiers which preserve the equivalence of IF prefixes.

Quantifier swapping. The prefixes

$$\Pi(Qu/U)(Q'v/V \cup u)\Pi'$$

and

$$\Pi(Q'v/V - u)(Qu/U \cup v)\Pi'$$

are equivalent. Notice that this operation changes the syntax of the prefix but does not affect the dependence and independence relations between quantiifers.

Dropping slash sets. The prefixes

$$(\forall v/V)\Pi \quad and \quad \forall v\Pi$$

are equivalent.

Emptying slash sets. The sentential prefixes

$$\Pi(Qv/V)\Pi'$$

and

$$\Pi Qv\Pi'$$

are equivalent whenever V contains variables that are quantified only by existential quantifiers in Π. Perhaps a surprising equivalence here is that between the prefixes $\exists x(\exists y/\{x\})$ and $\exists x\exists y$.

Notice what the first two operations achieve when they are combined: we can swap any universal quantifier $\forall x$ with any other quantifier (Qu/U) which occurs immediately at its right:

$$\Pi\forall x(Qu/U)\Pi'.$$

If $x \in U$, then by the swapping operation we obtain the equivalent prefix:

$$\Pi(Qu/U - x)(\forall x/u)\Pi'$$

which, after dropping the slashed set of the universal quantifier, is equivalent with

$$\Pi(Qu/U - x)\forall x\Pi'.$$

Example 7.6 The prefixes $\forall x(\exists y/\{x\})$ and $\exists y(\forall x/\{y\})$ are equivalent: take Π and Π' to be empty sequences, $Q = \forall, Q' = \exists$, U and V to be the empty set, $u = x$ and $v = y$. Now by applying the second operation, $\exists y(\forall x/\{y\})$ and $\exists y\forall x$ are equivalent.

Sevenster (2014) proves the following result (Lemma 14):

Lemma 7.3 *Let Π be an IF prefix and let Π´ be the IF prefix that is the result of*

- *swapping two quantifiers in Π (as in the first operation), or*
- *dropping the slash set of a universal quantifier in Π (as in the second operation), or*
- *emptying the slash set of a quantifier in Π that contains only existentially quantified variables (as in the last operation).*

Then, Π´ is primary, whenever Π is.

We prefer an example instead of the full proof.

Example 7.7 We consider the sentence φ_{MH} :

$$\forall x(\exists y/\{x\})\forall z(\exists t/\{x\})[x \neq z \wedge y \neq z \rightarrow x = t]$$

Notice first that the prefix of quantifiers is primary. We first swap the first two quantifiers to obtain the equivalent sentence:

$$\exists y(\forall x/\{y\})\forall z(\exists t/\{x\})[x \neq z \wedge y \neq z \rightarrow x = t]$$

Next we drop the slash set of the universal quantifier:

$$\exists y\forall x\forall z(\exists t/\{x\})[x \neq z \wedge y \neq z \rightarrow x = t].$$

We swap the two universal quantifiers:

$$\exists y\forall z\forall x(\exists t/\{x\})[x \neq z \wedge y \neq z \rightarrow x = t]$$

We swap the universal and existential quantifiers:

$$\exists y\forall z\exists t(\forall x/\{x\})[x \neq z \wedge y \neq z \rightarrow x = t]$$

Finally we drop the slashed set of the universal quantifier to obtain the first-order sentence

$$\exists y\forall z\exists t\forall x[x \neq z \wedge y \neq z \rightarrow x = t]$$

which is equivalent with the original one.

Lemma 7.4 (Sevenster, 2014) *Every sentential IF prefix $\Pi(\exists v/V)\Pi'$ in which Π is a FOL prefix is equivalent to a primary sentential IF prefix $\Sigma\Pi'$ in wich Σ is FOL.*

Proof Let Qx be a quantifier $Q = \forall$ such that it is closest to $(\exists v/V)$. If there is no such quantifier then all the quantifiers in the scope of which $(\exists v/V)$ are existential, which again means that all the variables in V are existentially quantified. We can then empty the slash set V to obtain the equivalent prefix $\Pi\exists v\Pi'$ which has the desired properties. Thus we shall assume that such a quantifier exists. We will show that by applying the rules above V can be emptied. There are two cases. Case 1: $x \notin V$; Case 2: $x \in V$.

Case 1. We have two subcases.

Case 1a. There is a universal quantifier $\forall y$ to the left of $\forall x$ such that $(\exists v/V)$ is independent of it, i.e. $y \in V$.

Case 1b. For all universal quantifiers $\forall y$ at the left of $\forall x$, $(\exists v/V)$ depends on them, i.e. $y \notin V$.

We prove both subcases.

Case 1a. We can represent the situation as in the picture

$$\ldots\forall y\ldots\forall x\underbrace{\quad}_{\exists}(\exists v/V)\Pi'$$

where $y \in V$ and $x \notin V$. Assume that a quantifier Qu appears between $\forall y$ and $\forall x$:

$$\ldots\forall y\ldots Qu\ldots\forall x\underbrace{\quad}_{\exists}(\exists v/V)\Pi'.$$

If Q is $\exists$, then there are two cases. a) $u \in V$. In this case the prefix is Henkin (see our example 8 above). b) $u \notin V$. In this case, the prefix is signaling (see our example 12 above). Both cases are excluded by the assumptions of the Lemma. So Qu must be universal, $\forall u$. We conclude that every quantifier between $\forall y$ and $\forall x$ must be universal:

$$\ldots \forall y \underbrace{\quad}_{\forall} \forall x \underbrace{\quad}_{\exists} (\exists v/V)\Pi'.$$

Let $\exists u$ be any existential quantifier between $\forall x$ and $(\exists v/V)$:

$$\ldots \forall y \underbrace{\quad}_{\forall} \forall x \underbrace{\quad}_{\exists} \exists u \underbrace{\quad}_{\exists} (\exists v/V)\Pi'.$$

Since the prefix is not signaling, we must have $u \notin V$. So we can swap $(\exists v/V)$ with any existential quantifier to its left until we reach $\forall x$:

$$\ldots \forall y \underbrace{\quad}_{\forall} \forall x(\exists v/V') \underbrace{\quad}_{\exists} (\exists u/v) \underbrace{\quad}_{\exists} \Pi'$$

where V' is the set of variables that remain after swapping. By Lemma 1, this is equivalent with

$$\ldots \forall y \underbrace{\quad}_{\forall} \forall x(\exists v/V') \underbrace{\quad}_{\exists} \exists u \underbrace{\quad}_{\exists} \Pi'$$

It follows by Lemma 1 that this prefix is primary.

Now we sap the universal quantifiers obtaining the equivalent prefix:

$$\ldots \forall x \underbrace{\quad}_{\forall} \forall y(\exists v/V') \underbrace{\quad}_{\exists} \exists u \underbrace{\quad}_{\exists} \Pi'$$

and sway $\forall y$ and $(\exists v/V')$ to obtain the equivalent prefix:

$$\ldots \forall x \ldots (\exists v/V' - y)\forall y \underbrace{\quad}_{\exists} \exists u \underbrace{\quad}_{\exists} \Pi'.$$

By Lemma 1 this prefix is primary. We have shown that we can drop y out of the slash set V.

Case 1b. Suppose $(\exists v/V)$ depends on all the universal quantifiers to the left of $\forall x$. Since $(\exists v/V)$ depends on $\forall x$, it follows that every variable in V is bounded by an existential quantifier. We then can empty V. By Lemma 2, the resulting prefix is primary.

Case 2. $x \in V$. If there is an existential quantifier $\exists u$ which appears between $\forall x$ and $(\exists v/V)$

$$\ldots \forall x \ldots \exists u \underbrace{\quad}_{\exists} (\exists v/V)\Pi'$$

then $u \in V$ because otherwise the prefix would be signaling. But then we can swap the two existential quantifiers obtaining the equivalent prefix:

$$\ldots \forall x(\exists v/V') \underbrace{\quad}_{\exists} (\exists u/v) \underbrace{\quad}_{\exists} \Pi'$$

and swap again:

$$\ldots(\exists v/V'-x)(\forall x/v)\ldots(\exists u/v)\ldots\Pi'$$

where V' is the set of variables that remains after swapping. We know that we can drop the slashed set of the universal quantifier, obtaining

$$\ldots(\exists v/V'-x)\forall x\ldots(\exists u/v)\ldots\Pi'.$$

Finally the slashed variable v is bounded by an existential quantifier, so by dropping it we get the desired result:

$$\ldots(\exists v/V'-x)\forall x\ldots\exists u\ldots\Pi'.$$ □

Theorem 7.5 (Sevenster 2014) *Every primary sentential IF prefix is equivalent to a sentential FOL prefix.*

PROOF Proof: We know that slashed sets may be dropped off from universal quantifiers. So what matters are only prefixes of the form

$$\Pi(\exists v/V)\Pi'.$$

But Lemma 2 shows that this prefix is equivalent to a FOL prefix. □

7.10 Playability of games: Perfect Recall

Informally a player has *action recall* if he always remembers his own moves, and *knowledge memory*, if he never forgets information he once knew. A player who has both action recall and knowledge memory is said to have *perfect recall.*

Notice that in a signaling sequence

$$\forall u\exists v(\exists w/\{u\})$$

the existential player does not have knowledge memory, for she knew the value $\forall u$ when making her first move, but not when she makes her second move. But she has action recall because she knows her first move when making her second move.

In a Henkin sequence

$$\forall x\exists u\forall y(\exists v/\{x,u\})$$

Eloise lacks both knowledge memory and action recall.

The following result connects a condition on the memory of the players (perfect recall) with expressive power:

Theorem 7.6 (Sevenster, 2006) *Every IF sentence for which Eloise has perfect recall is equivalent to an ordinary first-order sentence.*

A detailed proof may be found in Mann et al (2011), Theorem 6.2.3. As we saw, perfect recall subsumes both action recall and knowledge memory. The result above has been improved by Barbero (2014) who showed that action recall is the critical property: every sentential prefix with knowledge memory is equivalent to a sentential prefix of ordinary first-order quantifiers. We should be careful here. This result, however, does not characterize the IF prefixes that have a first-order equivalent, given that its converse does not hold. Here we have an example of an IF prefix where Eloise lacks knowledge memory (and lacks also action recall)

$$\forall x \exists y (\exists z / \{x, y\})$$

but which, after swapping the slashed quantifier twice, is equivalent to a first-order prefix

$$\exists z \forall x \exists y.$$

We already pointed out that Eloise lacks knowledge memory both in a signaling prefix and in a Henkin prefix. This suggests that a necessary and sufficient condition for an IF prefix to have an equivalent first order prefix:

Theorem 7.7 (Sevenster 2014) *Every sentential IF prefix is primary if and only if it is equivalent to a FOL prefix.*

PROOF We have already seen that every primary sentential IF prefix is equivalent to a sentential FOL prefix. For the converse, suppose that a prefix is not primary. Then it is either Henkin or signaling. If the former, then it extends one of the prefixes (1)-(6), and if the latter, it extends one of the prefixes (9)-(12). Theorem 12 in Sevenster (2014) shows that every sentential Henkin prefix expresses properties which are not first-order definable, and Theorem 23 shows the same thing for sentential signaling prefixes. □

Chapter 8

Probabilistic IF logic

8.1 Indeterminacy

We have analyzed two examples of indeterminate IF sentences, the Matching Pennies sentence φ_{MP}

$$\forall x(\exists y/\{x\})x = y$$

and the Inverted Matching Pennies sentence φ_{IMP}

$$\forall x(\exists y/\{x\})x \neq y.$$

Each of them is neither true nor false on every finite model with at least 2 elements (recall we are excluding models with 1 element). The indeterminacy of these sentences is nothing else than the indeterminacy of the underlying games of imperfect information.

There is a way to resolve the indeterminacy of IF sentences by following a suggestion due to Ajtai (reported in Blass and Gurevich 1986) in the context of branching quantifiers: we move to mixed strategies and then apply von Neumann minimax theorem to solve these games. In this approach, indeterminate IF sentences will receive intermediate truth-values on finite models:

- The value of

  $$\forall x(\exists y/\{x\})x = y$$

  is $\frac{1}{n}$.

- The value of

  $$\forall x(\exists y/\{x\})x \neq y$$

  is ${}^{n-1}/_{n}$.

Ajtai's suggestion has been implemented in Sevenster (2006) and generalised in Sevenster and Sandu (2010), Mann, Sandu and Sevenster (2011). Before describing it, let me give one more example of an indeterminate IF sentence.

Example 8.1 Recall the IF sentence φ_∞

$$\forall x \exists y (\exists z / \{x\})(c \neq y \wedge z = x)$$

which defines infinity. Let M be a model set which contains a finite number of elements, say n. Recall that φ_∞ is true in M if and only if there is a winning strategy for Eloise iff M is infinite. But M is finite, hence φ_∞ is not true in M. On the other side, suppose that φ_∞ is false in M. Then Abelard has a winning strategy in the underlying game, say τ. We notice that τ reduces to the choice of an individual $a \in M$ as a value for x. For τ to be winning Abelard must win all the plays where τ is used:

$$h_a = (\varphi_\infty, \varnothing, (x, a), (y, b), (z, d))$$

for every $b, c \in M$. That is, from all $b, d \in M$ we must have: $b = c^M$ or $a \neq d$. But this condition can be true only if M consisted of one single element. Given that we have excluded such models, then we conclude that φ_∞ is not false in M.

8.2 Strategic IF games: pure strategies

Recall the way in which we converted a semantical $G(\mathbb{M}, s, \varphi)$, where φ is a first-order formula, into a strategic game $\Gamma(\mathbb{M}, s, \varphi)$ that has the form $\Gamma = (S_\exists, S_\forall, \mu_\exists, \mu_\forall)$ where:

- $S_\exists$ is the set of strategies of Eloise in the semantic game $G(\mathbb{M}, s, \varphi)$
- $S_\forall$ is the set of strategies of Abelard in the semantic game $G(\mathbb{M}, s, \varphi)$
- $u_\exists$ and $u_\forall$ are the utility functions for Eloise and Abelard, respectively.

A strategic IF game $\Gamma(\mathbb{M}, s, \varphi)$ has exactly the same components, the only difference being that φ is now an IF formula. It is a 1-sum game, like its first-order relatives.

So a strategic IF game, like a strategic first-order game, can be displayed in the form of a matrix

	τ_1	τ_2	...	τ_n
σ_1	$u_\exists(\sigma_1, \tau_1)$	$u_\exists(\sigma_1, \tau_2)$	...	$u_\exists(\sigma_1, \tau_n)$
σ_2	$u_\exists(\sigma_2, \tau_1)$	$u_\exists(\sigma_2, \tau_2)$	...	$u_\exists(\sigma_2, \tau_n)$
$\vdots$	$\vdots$	$\vdots$	$\vdots$	$\vdots$
σ_m	$u_\exists(\sigma_m, \tau_1)$	$u_\exists(\sigma_m, \tau_2)$	...	$u_\exists(\sigma_m, \tau_n)$

where $S_\exists = \{\sigma_1, ..., \sigma_m\}$, $S_\forall = \{\tau_1, ..., \tau_n\}$. Here we have marked only the payoffs of Eloise, those of Abelard being automatically recoverable from them.

Example 8.2 We consider the game $\Gamma(\mathbb{M}, \varphi_{MP})$ where φ_{MP} is the sentence $\forall x(\exists y/\{x\})\, x = y$ and $\mathbb{M} = \{a, b\}$. Given the uniformity constraint, there are two strategies for Eloise: σ_1 : choose a as a value for y; and σ_2 : choose b for as a value for y. There are also two strategies for Abelard: τ_1 : choose a as a value for x; and τ_2 choose b as a value for x. The strategic form of the game yields the Matching Pennies game:

	a	b
a	$(1, 0)$	$(0, 1)$
b	$(0, 1)$	$(1, 0)$

Here is the strategic game form of the Inverted Matching Pennies game $\Gamma(\mathbb{M}, \varnothing, \varphi_{IMP})$ where φ_{IMP} is the sentence $\forall x(\exists y/\{x\})\, x \neq y$ and $\mathbb{M} = \{a, b\}$:

	a	b
a	$(0, 1)$	$(1, 0)$
b	$(1, 0)$	$(0, 1)$

As pointed out earlier, there is no equilibrium in the two games.

8.3 Strategic IF games: mixed strategies

In the case of the Matching Pennies game, the question is what is a rational behaviour for each of the two players, given that there is no equilibrium in pure strategies. We have pointed out earlier that one way out is for the players to randomize their strategies. Here we repeat some of the definitions:

- A mixed strategy μ for Eloise is a probability distribution over $S_\exists$, i.e. $\mu : S_\exists \to [0, 1]$ such that
$$\sum_{\sigma \in S_\exists} \mu(\sigma) = 1$$
- A mixed strategy ν for Abelard is a probability distribution over $S_\forall$, i.e. $\nu : S_\forall \to [0, 1]$ such that
$$\sum_{\tau \in S_\forall} \nu(\sigma) = 1$$

Given a mixed strategy μ for Eloise and a mixed strategy ν for Abelard, the expected utility for player i is given by

$$U_i(\mu, \nu) = \sum_{\sigma \in S_\exists} \sum_{\tau \in S_\forall} \mu(\sigma)\tau(\tau)u_i(\sigma, \tau)$$

8.4 Equilibrium semantics

We have seen that Von Neumann's Minimax Theorem guarantees that every IF game $\Gamma(\mathbb{M}, \varphi)$ has at least one equilibrium in mixed strategies. Given that

IF games are constant games, a corollary of this theorem tells us that any two equilibria in strategic IF games yield the same expected utility to the two players. We emphasize that these results hold only for finite structures. On infinite structures equilibria may not exist. For every IF sentence φ and finite structure M, the above results guarantee that we can talk about *the* expected utility returned to any of the players by any equilibrium of the strategic IF game $\Gamma(\mathbb{M}, \varphi)$.

Let φ be an IF sentence, and $\mathbb{M}$ a finite model. Suppose that (μ, ν) is an equilibria in the strategic IF game $\Gamma(\mathbb{M}, \varphi)$. Then we stipulate that *the value* of the sentence φ on $\mathbb{M}$, i.e. $v(\varphi, \mathbb{M})$ is the expected utility of the strategic IF game, $U_\exists(\mu, \nu)$:

$$v(\varphi, \mathbb{M}) = U_\exists(\mu, \nu).$$

When $v(\varphi, \mathbb{M}) = U_\exists(\mu, \nu) = \epsilon$, we shall write $\mathbb{M} \models_\epsilon \varphi$.

Equilibrium semantics turns out to be a conservative extension of game-theoretical semantics in the following sense:

Proposition 8.3 (Mann et al. 2011) For every IF sentence φ, finite model $\mathbb{M}$ and assignment s in M whose domain includes the free variables of φ we have:

(i) $\mathbb{M}, s \models^+_{GTS} \varphi$ if and only if $\mathbb{M}, s \models_1 \varphi$

(ii) $\mathbb{M}, s \models^-_{GTS} \psi$ if and only if $\mathbb{M}, s \models_0 \varphi$.

PROOF Assume first that σ is a winning strategy for Eloise in the the game $G(\mathbb{M}, s, \varphi)$. Then σ wins against any strategy τ of Abelard, that is, (*) $u_\exists(\sigma, \tau) = 1$, for every $\tau \in S_\forall$. Let ν be any mixed strategy of Abelard, i.e. $\nu \in \Delta(S_\forall)$. First, we claim that $U_\exists(\sigma, \nu) = 1$. To see this, recall that

$$U_\exists(\sigma, \nu) = \sum_{\tau \in S_\forall} \nu(\tau) u_\exists(\sigma, \tau)$$

which, in view of (*), yields:

$$U_\exists(\sigma, \nu) = \sum_{\tau \in S_\forall} \nu(\tau) = 1.$$

Let μ be the degenerate mixed strategy of Eloise which assigns to σ probability 1 (and 0 to all the other strategies in $S_\exists$). We claim that the pair (μ, ν) is an equilibrium. To show this we show that clauses (1)-(4) of Proposition 2.18 hold. Recall that

$$U_\exists(\mu, \nu) = \sum_{\sigma \in S_\exists} \mu(\sigma) U_\exists(\sigma, \nu).$$

But we just established that $U_\exists(\sigma, \nu) = 1$. Thus

$$U_\exists(\mu, \nu) = U_\exists(\sigma, \nu) = 1.$$

We conclude that clause (1) of Proposition 2.18 holds. For clause (2) we use the fact that (+) $u_\forall(\sigma, \tau) = 0$, for every $\tau \in S_\forall$. Let τ be any strategy in the support of ν. Recall that

$$U_\forall(\mu,\tau) = \sum_{\sigma\in S_\exists} \mu(\sigma)u_\forall(\sigma,\tau).$$

But given that for our fixed σ (+) holds, we get

$$U_\forall(\mu,\tau) = 0 = U_\forall(\mu,\nu).$$

This establishes our claim. Clauses (3) and (4) of Proposition 2.18 are immediate given that $U_\exists(\mu,\nu)$ is the maximal value in the game.

For the converse, assume that $\mathbb{M}, s \models_1 \varphi$, i.e. there is a pair (μ,ν) that is an equilibrium in the strategic IF game $\Gamma(\mathbb{M}, s, \varphi)$ such that $U_\exists(\sigma,\nu) = 1$. Let σ be any strategy in the support of μ, $\sigma \in S_\exists$. We claim that σ is a winning strategy of Eloise in the semantical game $G(\mathbb{M}, s, \varphi)$. To this effect, we are going to show that σ is winning against any strategy $\tau \in S_\forall$.

There are two cases: first we show that σ wins against any strategy in the support of ν; and, second, it also wins against any strategy τ' outside the support of ν.

For the first case, we use clause (1) of the Theorem in Chapter 3 for the identification of equilibria. We know that

$$U_\exists(\sigma,\nu) = U_\exists(\mu,\nu) = 1.$$

Recall that

$$U_\exists(\sigma,\nu) = \sum_{\tau\in S_\forall} \nu(\tau)u_\exists(\sigma,\tau)$$

But the pure strategies outside the support of ν have probability 0 and thus

$$(\#) \quad \textstyle\sum_{\tau\in S_\forall} \nu(\tau)u_\exists(\sigma,\tau) = \sum_{\tau\in T} \nu(\tau)u_\exists(\sigma,\tau)$$

where T is the set of strategies of Abelard in the support of ν. Suppose now, in order to derive a contradiction, that for some $\tau \in T$, $u_\exists(\sigma,\tau) = 0$. Then

$$\sum_{\tau\in T} \nu(\tau)u_\exists(\sigma,\tau) < 1$$

which implies

$$\sum_{\tau\in T} \nu(\tau) < 1.$$

But given (#), we have

$$\sum_{\tau\in S_\forall} \nu(\tau) < 1,$$

a contradiction. We conclude that $u_\exists(\sigma,\tau) = 1$, for every strategy τ in the support of ν.

Suppose now that τ is a strategy outside the support of ν. By clause (4) of the same Theorem we know that

$$U_\forall(\mu,\nu) \geq U_\forall(\mu,\tau)$$

But $U_\forall(\mu,\nu) = 0$, hence $U_\forall(\mu,\tau) = 0$, that is, $U_\exists(\mu,\tau) = 1$. Recall that

$$U_\exists(\mu,\tau) = \sum_{\sigma \in S_\exists} \mu(\sigma) u_\exists(\sigma,\tau)$$

which yields

$$U_\exists(\mu,\tau) = \sum_{\sigma \in S_\exists} \mu(\sigma) u_\exists(\sigma,\tau) = u_\exists(\sigma,\tau) = 1.$$

□

Thus σ wins against any strategy outside the support of ν. We conclude that σ is a winning strategy in the semantical game.

Note 8.4 Looking at the direction from right-to-left in the proof of the above Proposition, we observe that every strategy in the support of μ is winning. That is, (σ,ν) is an equilibrium for every strategy σ in the support of μ. This shows that randomizing over winning strategies does not improve the expected utility.

Example 8.5 We applyTheorem in Chapter 3 to the strategic games $\Gamma(\mathbb{M},\varphi_{MP})$ and $\Gamma(\mathbb{M},\varphi_{IMP})$, where $\mathbb{M} = \{a,b\}$, φ_{MP} is $\forall x(\exists y/\{x\}\, x = y$ and φ_{IMP} is $\forall x(\exists y/\{x\}\, x \neq y$. We represented them in our earlier example as

	a	b
a	$(1,0)$	$(0,1)$
b	$(0,1)$	$(1,0)$

	a	b
a	$(0,1)$	$(1,0)$
b	$(1,0)$	$(0,1)$

Let μ and ν be uniform probability distributions over $\{a,b\}$, i.e. $\mu(a) = \mu(b) = 1/2 = \nu(a) = \nu(b)$. The pair (μ,ν) is an equilibrium in both games. For the Matching Pennies game:

$$U_\exists(a,\nu) = \textstyle\sum_{\tau \in S_\forall} \nu(\tau) u_\exists(a,\tau) = (1/2 \times 1) + (1/2 \times 0) = 1/2.$$
$$U_\exists(b,\nu) = \textstyle\sum_{\tau \in S_\forall} \nu(\tau) u_\exists(b,\tau) = (1/2 \times 0) + (1/2 \times 1) = 1/2.$$

Thus clause (1) of the Theorem is satisfied. Clause (2) is proved in a similar way. Clauses (3) and (4) are trivially satisfied. The computations for the inverted Matching Pennies game are symmetrical. Notice that when the cardinality of M increases in the Matching Pennies game

	1	2	...	n
1	$(1,0)$	$(0,1)$	...	$(0,1)$
2	$(0,1)$	$(1,0)$	...	$(0,1)$
⋮	⋮	⋮	⋮	⋮
n	$(0,1)$	$(0,1)$	...	$(1,0)$

and in the Inverted Matching Pennies game

	1	2	...	n
1	(0,1)	(1,0)	...	(1,0)
2	(1,0)	(0,1)	...	(1,0)
$\vdots$	(1,0)	(1,0)	...	$\vdots$
n	(1,0)	(1,0)	...	(0,1)

then the value of φ_{MP} on $\mathbb{M}$ is $^1/_n$ and that of φ_{IMP} is $^{n-1}/_n$.

8.5 Equilibrium semantics: Case studies

We shall give here a couple of examples which illustrates the interaction between game-theoretical and logical methods. Both examples illustrate the phenomenon of signaling discussed in chapter 5. Actually the first example shows how we can define Lewis signaling, Sender-Receiver games in IF logic. The second example is the sentence defining infinity in IF logic. It turns out that this sentence and the underlying game are also an example of signaling, so it should come as no surprise that the methodology for treating the two cases is the same: even if the two examples are apparently far removed from each other, in both of them the two existential players have to coordinate themselves on a number of "signals" (messages). We shall use, in addition to Theorem 4.3 in Chapter 4, two other game-theoretical results which will help us to reduce the size of the relevant game.

In a game, given the assumption of the rationality of the players being common knowledge, there may be strategies which are not worth playing, because some other strategies ensure a better payoff. We start with a couple of definitions. We first recall the notion of weakly dominated strategy.

Fix a strategic IF game Γ. We say that a pure strategy σ of Eloise, $\sigma \in S_\exists$, is *weakly dominated* by another pure strategy $\sigma' \in S_\exists$ if

- For all pure strategies τ of Abelard, $u_\exists(\sigma,\tau) \geq u_\exists(\sigma',\tau)$, and
- There is at least one pure strategy τ' of Abelard such that $u_\exists(\sigma,\tau') > u_\exists(\sigma',\tau')$

The definition for a strategy of Abelard being weakly dominated by another strategy is completely similar. It is reasonable to expect that a weakly dominated strategy be assigned probability 0 by a player in a mixed strategy equilibrium.

There are also strategies which are undistinguishable in the sense that they return the same payoff to a player.

- For $\sigma, \sigma' \in S_\exists$, we say that σ' is *payoff equivalent* to σ if for every $\tau \in S_\forall$: $u_\exists(\sigma',\tau) = u_\exists(\sigma,\tau)$.

- There is at least one pure strategy τ' of Abelard such that $u_\exists(\sigma, \tau') > u_\exists(\sigma', \tau')$

A similar definition holds for Abelard.

The following result is well known in game theory. For the proof see Osborne (2004) and Mann et al (2011), proposition 7.21.

Proposition 8.6 Let $\Gamma(\mathbb{M}, \varphi)$ be a strategic IF game. Then Γ has an equilibrium $(\sigma_\exists, \sigma_\forall)$ such that for each player p none of the strategies in the support of σ_p is weakly dominated in Γ.

We list another result which allows the elimination of payoff equivalent strategies.

Proposition 8.7 Let $\Gamma(\mathbb{M}, \varphi)$ be a strategic IF game. Then Γ has equilibrium $(\sigma_\exists, \sigma_\forall)$ such that for each player p there are no strategies in the support of σ_p which are payoff equivalent.

PROOF Mann et al (Proposition 7.22) proves this result in the context of IF logic. □

These results tell us that when looking for equilibria in a strategic IF game Γ, we may look for an equilibria in a smaller game in which weakly dominated and payoff equivalent strategies have been removed.

We now apply the framework developed so far to two case studies. Both of them involve the signaling prefix $\forall x \exists z (\exists y / \{x\})$.

8.6 Lewis's signaling games in IF logic

We follow Barbero and Sandu (2014). We express Lewis signaling systems in IF logic using the following IF sentence:

$$\varphi_{sig} = \forall x \exists y (\exists z / \{x\}) \{(W(x) \rightarrow (\Sigma(y) \wedge R(z) \wedge z = x))\}$$

The symbolism is self-explanatory: W stands for the set of states, Σ for the set of signals, and R for the set of responses.

We consider the class of models of the form

$$\mathbb{M} = (M, W^M, \Sigma^M, R^M)$$

where

$$M = \{s_1, ..., s_n, t_1, ..., t_m\}$$
$$W^M = R^M = \{s_1, ..., s_n\}$$
$$\Sigma^M = \{t_1, ..., t_m\}.$$

This sentence says that for every state, the first existential player (the Sender) chooses a message and the second existential player (the Receiver) chooses a response, which is also a state, identical to the state chosen by Nature. We notice that the role performed by Nature in our earlier Sender-Receiver Bayesian games of incomplete information is now performed by Abelard; and the role of a response being the "best action" for a given state is now taken by the Receiver identifying the relevant state. In addition, Nature (Abelard) has a strategic role. φ_{sig} is determinate (true) in all models $\mathbb{M} = (M, W^{\mathbb{M}}, \Sigma^{\mathbb{M}}, R^{\mathbb{M}})$ where the number m of signals equals the number of states, that is, $\mid W^{\mathbb{M}} \mid = n = \mid \Sigma^{\mathbb{M}} \mid = m$. It is false when there are no signals, i.e. $\Sigma^{\mathbb{M}} = \varnothing$. But when the number m of signals is strictly less than the number of states, that is, $0 < m < n$, then φ_{sig} is indeterminate. Let us show this.

Suppose, for a contradiction, that $\mathbb{M} \models^{+} \varphi_{sig}$, that is, there is a winning strategy for Eloise in the game $G(M, \varnothing, \varphi_{sig})$. As in the case of sentence φ_{∞} defining infinity, σ may be associated with a pair of functions (f, g), where $f : M \to M$ is the function which yields a value for y, such that

$$\sigma(\varphi_{\infty}, \varnothing, (x, a)) = (y, f(a))$$

for every $a \in M$; and $g : M \to M$ is the function which yields a value for z, such that

$$\sigma(\varphi_{\infty}, \varnothing, (x, a), (y, b)) = (z, g(b)),$$

for every $a, b \in M$. Given that σ is winning Eloise must win every play in which she uses σ, that is

$$\texttt{If} \quad a \in W^M \quad \texttt{then} \quad f(a) \in \Sigma^M \quad \texttt{and} \quad g(f(a)) = a \quad \texttt{and} \quad g(f(a)) \in R^M$$

for every $a \in M$. Given that then number of signals is less than the number of states, i.e. $m < n$, and every state $s \in W^M$ is associated with a signal, i.e., $f(s) \in \Sigma^M$, there must exist distinct states s_i, s_j which are mapped onto the same signal: $f(s_i) = f(s_j) = t$, for some $t \in \Sigma^M$. But this contradicts the fact that f is an injection, that is, $g(f(s_i)) = s_i$ and $g(f(s_j)) = s_j$, a contradiction.

Suppose now that $\mathbb{M} \models^{-} \varphi_{sig}$, that is, there is a winning strategy for Abelard in the game $G(\mathbb{M}, \varnothing, \varphi_{sig})$. Winning strategies for Abelard reduce to the choice of an individual in the universe. For this reason we shall represent them as indexed with the individual they choose. Thus τ_a is the strategy of Abelard such that

$$\tau_a(\varphi_{sig}, \varnothing) = (x, a).$$

Notice that a strategy τ_a is winning iff

$$a \in W^M \quad \texttt{and} \quad (b \notin \Sigma^M \quad \texttt{or} \quad c \notin R^M \quad \texttt{or} \quad c \neq a)$$

for all $b, c \in M$.

So suppose τ_a is a winning strategy for Abelard. Hence a satisfies the condition above for all $b, c \in M$. Pick up $b = t_i \in \Sigma^M$ and $c = a$. Then the second conjunct of the condition is violated, a contradiction. Note that for φ_{sig} to be indeterminate, M has to have at least three elements: at least one signal and at least two states.

We now compute the value of φ_{sig} in M according to the equilibrum semantics.

The strategic game has the form

$$\Gamma(\mathbb{M}, \varphi_{sig}) = (S_\exists, S_\forall, u_\exists, u_\forall)$$

where $S_\exists$, the set of strategies of Eloise, will be taken to consist of pairs of functions (f, g) as indicated above, and $S_\forall$, the set of strategies of Abelard, will be taken to have the form $S_\forall = \{\tau_a : a \in M\}$.

The strategic game may be displayed in the matrix form:

	τ_{s_1}	$\cdots$	τ_{s_n}	τ_{t_1}	$\cdots$	τ_{t_m}
(f_1, g_1)				$(1,0)$	$\cdots$	$(1,0)$
(f_2, g_2)				$(1,0)$	$\cdots$	$(1,0)$
$\vdots$				$\vdots$	$\vdots$	$\vdots$
(f_p, g_p)				$(1,0)$	$\cdots$	$(1,0)$

Notice that any two strategies τ_{t_i}, τ_{t_j} are payoff equivalent for Abelard, that is:

- $u_\forall((f, g), \tau_{t_i}) = u_\forall((f, g), \tau_{t_j})$ for any $(f, g) \in S_\exists$.

Let $B \subseteq W^M$ such that $\mid B \mid = m$. Let S_B be the set of pair of functions (f, g) in the set $S_\exists$ which fulfill the condition:

A For all $r_1, ..., r_m$ in B we have

$$f(r_1) \in \Sigma^M \wedge ... \wedge f(r_m) \in \Sigma^M$$

and

$$g(f(r_1)) = r_1 \wedge ... \wedge g(f(r_m)) = s_m.$$

Notice that this condition says, essentially, that the two existential quantifiers (Sender and Receiver) achieve "coordination" on the states in B.

Given that $m < n$, the set S_B is nonempty. Here is an example of such a pair (f^*, g^*) for $B = \{s_1, ..., s_m\}$ (we do not list the values of f^* and g^* for the states not in B):

$$\begin{array}{ccccc} s_1 & \stackrel{f*}{\to} & t_1 & \stackrel{g^*}{\to} & s_1 \\ \vdots & \vdots & \vdots & \vdots & \vdots \\ s_m & \stackrel{f^*}{\to} & t_m & \stackrel{g^*}{\to} & s_m \\ \vdots & \vdots & \vdots & \vdots & \vdots \end{array}$$

Obviously the payoff of the outcomes where Eloise plays (f^*, g^*) and Abelard plays any of the strategies τ_{s_i} is $(1, 0)$ whenever $i \leq m$ and $(0, 1)$ for $i > m$:

	τ_{s_1}	$\cdots$	τ_{s_m}	$\cdots$	τ_{s_n}	$\cdots$
(f^*, g^*)	$(1,0)$	$(1,0)$	$(1,0)$	$(0,1)$	$(0,1)$	$(1,0)$

But then two things follow:

- Every strategy τ_{s_i} weakly dominates any strategy τ_{t_j} for Abelard, that is, for every $(f, g) \in S_\exists$

$$u_\forall((f,g),\tau_{s_i}) \geq u_\forall((f,g),\tau_{t_j})$$

and for some $(f, g) \in S_\exists$

$$u_\forall((f,g),\tau_{s_i}) > u_\forall((f,g),\tau_{t_j}).$$

- Any pairs (f, g) and (f', g') in the set S_B are payoff equivalent for Eloise, i.e. for every $\tau \in S_\forall$ we have

$$u_\exists((f,g),\tau) = u_\exists((f',g'),\tau).$$

To see that the second condition holds, notice that

$$u_\exists((f,g),\tau_s) = u_\exists((f,g),\tau_s) = 1$$

for all τ_s such that $s \in B$; and

$$u_\exists((f,g),\tau_s) = u_\exists((f,g),\tau_s) = 0$$

for all τ_s such that $s \in \{t_1, ..., t_m\}$; and

$$u_\exists((f,g),\tau_s) = u_\exists((f,g),\tau_s) = 0$$

for all τ_s such that $W^M - B$.

Then by Proposition 8.6, we remove the weakly dominated strategies reducing the game to the smaller one:

	τ_{s_1}	$\cdots$	τ_{s_n}
(f_1, g_1)			
(f_2, g_2)			
$\vdots$			
(f_p, g_p)			

which has the same value. By 8.7, we may assume that for each $B \subseteq W^M$ such that $\mid B \mid = m$ there is exactly one (f, g) such that $(f, g) \in S_M$. Let us represent each such pair as $(f, g)_B$.

We claim that every strategy of Eloise in the smaller game is weakly dominated by some strategy $(f, g)_B$. The following informal argument should suffice to prove this claim.

If a pair (f, g) is in none of the sets S_B then the functions f and g "coordinate" on less then m states. Suppose they coordinate on r states, $r < m$. Let the states for which coordination is successful be $B' = \{s_1, ..., s_r\}$. Then $u_\exists((f, g), s_i) = 1$ for every $s_i \in B'$ and 0 for every $s_i \in W^M - B'$. But now let B such that $B' \subseteq B$ and $\mid B \mid = m$. It is easy to see that the pair $(f', g')_B$ weakly dominates (f, g).

By the theorem above we can further reduce the game to the smaller format $S \times T$ where $S = \{(f,g)_B : B \subseteq W^M, \mid B \mid = m\}$ and $T = \{\tau_c : c \in W^S\}$. Each of the strategies in S has in its row m times the payoff $(1,0)$ and $n-m$ times the payoff $(0,1)$. Let μ be the uniform probability distribution with support the sets S_B and ν the uniform probability distribution with support S^M. We claim that the pair (μ, ν) is an equilibrium with value ${}^m/_n$. The cardinality of the set of pairs of functions in the support of μ will be computed below. However the expected utility of the players does not depend on it. Suppose it is p. Then the probability of each (f_i, g_i) is ${}^1/_m$ and the probability of each s_i is ${}^1/_n$. We have:

$$U_\exists(\mu,\nu) = p \times {}^1/_p \times {}^1/_n \times m = {}^m/_n$$

and thus $U_\exists(\mu,\nu) = 1 - {}^m/_n$.

It still remains to show that (μ,ν) is an equilibrium in the smaller game. Clauses (3)and (4) of the Theorem 4.3 are trivially satisfied. For clause (1), let $(f,g) \in S$. Given that Eloise receives m times the payoff 1, we have:

$$U_\exists((f,g),\nu) = \sum_{\tau \in T} \nu(\tau) u_\exists((f,g),\tau) = {}^1/_n \times m = {}^m/_n.$$

We are done. Clause (2) is more tedious. Let $\tau_a \in T$. We know that

$$U_\forall(\mu,\tau_a) = \sum_{(f,g)\in S} \mu(f,g) u_\forall((f,g),\tau).$$

But in order to be able to compute $U_\forall(\mu,\tau_a)$ we need to know two more things: a) the value of $\mu(f,g)$; and b) how many times Abelard gets the payoff 1.

For (a), $\mu(f,g) = {}^1/_p$, where p is the cardinality of S. So we need to know how many subsets with m elments W^M has. A textbook result in combinatorics tells us that

$$p = \binom{n}{m} = {}^{n!}/_{m!(n-m)!}.$$

As for (b) we shall calculate it in the following way. When playing τ_a, Abelard will get payoff 1 whenever Eloise plays a strategy $(f,g)_B$ which is not winning against τ_a, that is, whenever $a \notin B$. So we need to know first how many subsets B of W^M of cardinality m there are which do not contain a. We will find that out if we knew how many subsets B of W^M of cardinality m there are which do contain a. Their number is

$$q = \binom{n-1}{m-1} = {}^{(n-1)!}/_{(m-1)!(n-m)!}$$

Thus

$$\begin{aligned} U_\forall(\mu,\tau_a) &= \textstyle\sum_{(f,g)\in S} \mu(f,g) u_\forall((f,g),\tau) \\ &= {}^1/_p \times (p-q) = 1 - {}^q/_p = \\ 1 - \binom{n-1}{m-1} &\Big/ \binom{n}{m} = 1 - {}^m/_n. \end{aligned}$$

We conclude that (μ, ν) is an equilibrium.

The definability of Lewis signaling games in IF logic provides us with a new proof of the following theorem.

Theorem 8.1 (Barbero and Sandu, 2014) *For every integers m, n such that $0 \leq m < n$ there is a structure $\mathbb{M} = (M, S^{\mathbb{M}}, \Sigma^{\mathbb{M}}, R^{\mathbb{M}})$ such that the value of φ_{sig} in this structure is $^m/_n$.*

8.7 Infinity on finite structures

The material in this section is based on Mann et al (2011) but we present it in a slighlty different way to emphasize the similarity with the proof in the previous section. Recall the sentence φ_∞ which defines infinity in IF logic:

$$\forall x \exists y (\exists z / \{x\}) \left[y \neq c \wedge z = x \right].$$

Let M be a finite structure which interprets c. We have seen (section 8.6.) that φ_∞ is indeterminate on M. It is then natural to ask what is its probabilistic value on M.

We consider the strategic game $\Gamma(\mathbb{M}, \varphi_\infty) = (S_\exists, S_\forall, \mu_\exists, \mu_\forall)$. We have seen in our earlier analysis of the semantic game of this sentence that we can take $S_\forall$ to consist of elements of the form τ_a such that

$$\tau_a(\varphi_\infty, \varnothing) = (x, a).$$

We also saw how to associate with each $\sigma \in S_\exists$ a pair of functions (f, g), where $f : M \to M$ is the function which yields a value for y, such that for every $a \in M$

$$\sigma(\varphi_\infty, \varnothing, (x, a)) = (y, f(a))$$

and $g : M \to M$ is the function which yields a value for z, such that for every $a, b \in M$

$$\sigma(\varphi_\infty, \varnothing, (x, a), (y, b)) = (z, g(b)).$$

Notice that if $\sigma \in S_\exists$ and $\tau_a \in S_\forall$ then σ wins against τ_a (i.e. $u_\exists(\sigma, \tau_a) = 1$) if and only if

$$g(f(a)) = a \quad \text{and} \quad f(a) \neq c^M$$

where (f, g) is the pair associated with σ. We know that there is no $\sigma \in S_\exists$ which wins against every $\tau_a \in S_\forall$. For if that were the case, the left conjunct above would require f to be an injection, whose range, by the requirement imposed by the right conjunct, would also have to be distinct from the element c^M. But that can't be the case in a finite universe.

The present game, like the previous one, is a signaling game. Although the two existential players cannot "coordinate" on the whole M, that is, although they cannot win against all the strategies in $S_\forall$, they can coordinate on many of them. Here is an example. Let M be a five element structure, $M = \{a_1, ..., a_5\}$, where $c^M = a_3$. Let (f, g) be the pair of functions such that $f = g$ and f is

$$\begin{array}{ccc} a_1 & \to & a_1 \\ a_2 & \to & a_2 \\ a_3 & \to & a_4 \\ a_4 & \to & a_3 \\ a_5 & \to & a_3 \end{array}$$

Notice that (f, g) wins against τ_a where $a \in \{a_1, a_2, a_3\}$ and looses against $a \in \{a_4, a_5\}$. In order to reduce the game, we shall follow the same idea as in the analysis of the signaling game in the previous section: to find pairs of strategies (f, g) which are winning against a "maximal" number of elements of the universe (strategies of Abelard) and to show that every other strategy (f, g) is weakly dominated by one of those. The set of strategies we ended up consisted of pairs (f, g), each associated with a set $B \subseteq M$ of cardinality m: (f, g) was winning against any element in B.

In the present case, from what φ_∞"says" and from the fact that the universe is finite we can see that the best result a strategy pair (f, g) can achieve is to win against all elements of the universe, except one: c^M. So it should not come as a surprise that the set of strategies (f, g) we will end up with will be associated with singletons $\{a\}\subseteq M$, this time the idea being that (f, g) wins against all elements of M except a.

Let $A \subseteq M$. We stipulate that S_A is the set of strategies (f, g) that win against any element outside A, i.e.,

$$g(f(a)) = a \quad \texttt{and} \quad f(a) \neq c^M \quad \texttt{iff} \quad a \notin A$$

Once again, for $(f, g) \in S_A$, (f, g) wins against Abelard's strategy τ_a if and only if $a \notin A$. Thus in our example above, $(f, g) \in S_{\{a_4, a_5\}}$. Notice also that, given that Eloise has no winning strategy, S_A is empty for $A = \varnothing$; and given that Abelard has no winning strategy, S_A is also empty for $A = M$.

Notice also that every strategy σ (or more exactly its corelated pair (f, g)) belongs to some non-empty set $A \subseteq M : A = M - B$, where B is the set of elements a such that σ wins against τ_a.

Let $A \subseteq M$. We claim that all the strategies in S_A are payoff equivalent. For let $\sigma, \sigma' \in S_A$ and let τ_a be an arbitrary strategy in $S_\forall$. By the construction of S_A

$$\begin{array}{lll} u_\exists(\sigma, \tau_a) = 1 & \texttt{iff} & a \notin A \\ u_\exists(\sigma', \tau_a) = 1 & \texttt{iff} & a \notin A. \end{array}$$

Hence $u_\exists(\sigma, \tau_a) = u_\exists(\sigma', \tau_a)$ for any $\tau_a \in S_\forall$. We now show that for any $a \in M$ the set $S_{\{a\}}$ is nonempty.

To see this, fix $a \in M$. We let $f, g : M \to M$ such that $f = g$ be the function

$$f(z) = \begin{cases} a & if\ z = c^M \\ c^M & if\ z = a \\ z & otherwise. \end{cases}$$

To give an example, let M be a five element structure from our earlier example. To show that $S_{\{a_1\}}$ is non-empty we let $f = g$ be the function:

$$\begin{array}{lcl} a_1 & \to & a_3 \\ a_2 & \to & a_2 \\ a_3 & \to & a_1 \\ a_4 & \to & a_4 \\ a_5 & \to & a_5 \end{array}$$

We return to the general case. It is easy to see from the way the functions f and g have been defined that $g(f(z)) = z$ for every $z \in M$. Notice also that $f(z) = c^M$ if and only if $z = a$. Hence $(f, g) \in S_{\{a\}}$.

Let S' be the set of strategies of Eloise which collect one representative strategy from each $S_{\{a\}}$, $a \in M$. We claim that every strategy σ in $S_\exists$ is weakly dominated or payoff equivalent to some strategy σ' in S'. To show this, fix σ and let $B \subseteq M$ a nonempty set such that $\sigma \subseteq S_B$. Let τ_c be an arbitrary strategy of Abelard. Suppose that σ wins against τ_c, i.e., $u_\exists(\sigma, \tau_c) = 1$. Hence, by the construction of B, $c \notin B$. Then for any singleton $\{a\} \subseteq B$ we have: $c \notin \{a\}$, (i.e. $c \neq a$) for otherwise $c \in B$, a contradiction. Thus for any strategy σ' in $S' \cap S_{\{a\}}$ we have $u_\exists(\sigma', \tau_c) = 1$.

By Proposition 8.6. the value of the initial game is the same as the value of the smaller game where $S_\exists = S'$ and $S_\forall = \{\tau_b : b \in M\}$. Obviously we can write $S' = \{\sigma_a : a \in M\}$ where σ_a is the strategy from the set $S_{\{a\}}$. Consider the situation in which Eloise plays σ_a and Abelard palys τ_b. Eloise wins iff $a \neq b$. It follows that the payoff matrix of the new game is identical to the payoff matrix of the Inverted Matching Pennies. Thus the value of φ_∞ on M is $^{n-1}/_n$. As the cardinality of M grows to infinity, the value of φ_∞ approaches 1.

It is interesting to notice that the value of φ_∞ on finite structures is the same as the value of the Inverted Matching Pennies sentence

$$\varphi_{IMP} = \forall x(\exists y/\{x\})x \neq y.$$

8.8 What kind of probabilistic logic is Probabilistic IF logic?

The material in this section is based on Parvu et al (forthcoming). For any IF sentence φ, finite model $\mathbb{M}$ and assignment s in $\mathbb{M}$ whode domain inlcudes the free variable of φ, let $NE(\varphi, s, \mathbb{M})$ denote the value of the IF sentence φ on the $\mathbb{M}$. When s is empty we write $NE(\varphi, s, \mathbb{M})$. Mann, Sandu and Sevenster (2011) prove the following properties of $NE(\varphi, s, \mathbb{M})$:

P1 $\quad NE(\varphi \vee \psi, s, \mathbb{M}) = max\,\{NE(\varphi, s, \mathbb{M}), NE(\psi, s, \mathbb{M})\}$

P2 $\quad NE(\varphi \wedge \psi, s, \mathbb{M}) = min\,\{NE(\varphi, s, \mathbb{M}), NE(\psi, s, \mathbb{M})\}$

P3 $\quad NE(\neg\varphi, s, \mathbb{M}) = 1 - NE(\varphi, s, \mathbb{M})$.

P4. $NE(\exists x\varphi, s, \mathbb{M}) = max\,\{NE(\varphi, s(x/a), \mathbb{M}) : a \in M\}$

P5. $NE(\forall x\varphi, s, \mathbb{M}) = max\,\{NE(\varphi, s(x/a), \mathbb{M}) : a \in M\}$

On the basis of these properties it is easy to see that IF probabilities obey Kolmogorov axioms:

1. $NE(\varphi, \mathbb{M}) \geq 0$
2. $NE(\varphi, \mathbb{M}) + NE(\neg\varphi, \mathbb{M}) = 1$
3. $NE(\varphi, \mathbb{M}) + NE(\psi, \mathbb{M}) \geq NE(\varphi \vee \psi, \mathbb{M})$
4. $NE(\varphi \wedge \psi, \mathbb{M}) = 0 \rightarrow NE(\varphi, \mathbb{M}) + NE(\psi, \mathbb{M}) = NE(\varphi \vee \psi, \mathbb{M})$

(1) follows from the basic definitions. (2) follows from (P3). (3) follows from (P1) and (1). For (4), suppose that $NE(\varphi \wedge \psi, \mathbb{M}) = 0$ is true in a model $\mathbb{M}$, i.e., $\mathbb{M} \models_0 \varphi \wedge \psi$. Then by the conservativity of the equilibrium semantics over classical logic (Proposition 8.3.), $\mathbb{M} \models^-_{GTS} \varphi \wedge \psi$. From the game-theoretical semantics of IF logic, it follows that either $\mathbb{M} \models^-_{GTS} \varphi$ or $\mathbb{M} \models^-_{GTS} \psi$. If the former, then we infer using the conservativity property that $NE(\varphi, \mathbb{M}) = 0$. But then, using (P1) we get:

$$\begin{aligned} NE(\varphi \vee \psi, \mathbb{M}) &= max\,\{NE(\varphi, \mathbb{M}), NE(\psi, \mathbb{M})\} \\ &= NE(\psi, \mathbb{M}) \\ &= NE(\varphi, \mathbb{M}) + NE(\psi, \mathbb{M}). \end{aligned}$$

The case $\mathbb{M} \models^-_{GTS} \psi$ is similar.

8.8.1 Probabilistic semantics: degrees of rational belief

There are various ways to combine logic and probabilities. We shortly review here a probabilistic semantics for the language of predicate logic (Leblanc 1994). Let L be a first-order language. The logical primitives of L are negation, conjunction and the univeral quantifier; disjunction, implication and the existential quantifier are defined in the standard way. I will focus here only on the propositional part of L. We associate with L a probability function P that takes single sentences of L as arguments (in the same way as truth-value functions do) and yield real numbers from the entire interval $[0, 1]$ as values. The classical truth values 1 and 0 are assimilated to the endpoints of the interval $[0, 1]$. A sentence's probability value is seen as a measure of its (un)certainty, with 0 representing maximal uncertainty and 1 maximal certainty. A probability function P must obey the following constraints:

C1. $P(\varphi) \geq 0$, for all $\varphi \in L$

C2. $P(\neg(\varphi \wedge \neg\varphi)) = 1$

C3. $P(A) = P(A \wedge B) + P(A \wedge \neg B)$

C4. $P(A) \leq P(A \wedge A)$

C5. $P(A \wedge B) \leq P(B \wedge A)$

C6. $P(A \wedge (B \wedge C)) \leq P^{+}(A \wedge B) \wedge C)$.

C1 and C2 are the counterparts of the axioms known as non-negativity and unit normalization in Kolmogorov (1933). C3-C6 are borrowed from Popper (1955).

In considering other alternative constraints, Leblanc (1994) discusses two additional principles mentioned by Kolmogorov as Additivity and Continuity. The logical counterpart of the latter is relevant for infinite disjunctions and thus does not apply here. And the logical counterpart of the former is:

C7. If two sentences φ and ψ of L are logically incompatible in the standard sense, i.e., $\models \neg(\varphi \wedge \psi)$, then $P(\varphi \vee \psi) = P(\varphi) + P(\psi)$.

In addition, Leblanc also mentions the following principle of *Substitutivity*

C8. If φ and ψ are logically equivalent in the standard sense (i.e., $\models \varphi \leftrightarrow \psi$), then $P(\varphi) = P(\psi)$

that he finds quite natural in the context of logic. He shows (Theorem 4.40) that C1-C2 together with C7-C8 pick up the same probability function as C1-C6.

We are also told that Popper (1955) uses the following three principles which replace C1-C2:

C9. $P(\varphi \wedge \psi) \leq P(\varphi)$ (Monotonicity)

C10. There are at least two distinct statements φ and ψ of L such that $P(\varphi) \neq P(\psi)$ (Existence).

C11. For each statement φ of L there is a statement ψ of L such that $P(\varphi) \leq P(\psi)$ and $P(\varphi \wedge \psi) = P(\varphi) \times P(\psi)$.

Leblanc (1982c) shows that C3-C6 and C9-C11 pick out the same probability function as C1-C6.

Among the theorems of (C1)-(C6) one finds:

- the commutativity, associativity and idempocy (i.e., $P(\varphi) = P(\varphi \wedge \varphi)$) of conjunction,
- the probility of a conjunction is less or equal to the probability of each conjunct,
- contradictions $(\varphi \wedge \neg\varphi)$ get probability 0 and $\neg\varphi$ gets probability $1 - P(\varphi)$,
- $P(\varphi \rightarrow \psi) = 1$ iff $P(\varphi \wedge \neg\psi) = 0$ iff $P(\varphi) = P(\varphi \wedge \psi)$,
- the principle of substitutivity (C9): if $\varphi \leftrightarrow \psi$ is a theorem of propositional logic, then $P(\varphi) = P(\psi)$.

8.8.2 Probabilistic semantics: statistical knowledge (relative frequency)

The main ideas have been introduced in Bacchus (1990). In the preceding case the syntax of the underlying propositional and predicate logic languages remained standard. In the present case, both the syntax of predicate logic and its semantics are enriched with probabilistic features. The basic logical primitives are negation, disjunction and the existential quantifier. In addition to the standard syntactical clauses for first-order formulas, we have a new syntactical clause:

- When φ is a standard first-order formula and q is a rational number in the interval $[0, 1]$, then $Px(\varphi) \geq q$ is a formula with the intended interpretation: "the probability of selecting an x such that x satisfies φ is at least q". $Px(\varphi) = q$ is an abbreviation of

$$Px(\varphi) \geq q \wedge Px(\varphi) \leq q.$$

Here every free occurrence of x in φ is bound by the operator Px.

Unlike in the previous framework, we keep models (and the notion of reference), but we enrich them with a probability distribution over the individuals in the universe. That is, models are now triples $M = (D, I, P)$, where the domain of discourse D and the interpretation function I are standard, and P is a probability function such that $\Sigma_{d \in D} P(d) = 1$. Thus probabilities are associated in the first place with the individuals of the universe. As a result, each term of the form $Px(\varphi)$ receives a probabilistic value in a way to be sketched below.

First, ordinary first-order formulas are interpreted in the model $M = (D, I, P)$ relative to an assignment g in the standard way.

Second, we have one additional clause:

$$M, g \models Px(\varphi) \geq q \quad \texttt{iff} \quad \Sigma_{d \in D: M, g[x \to d] \models \varphi} \geq q.$$

That is, the probability of selecting an x that satisfies φ is the sum of the probabilities of the elements in the universe that satisfy φ.

To get a better grasp on what is going on, here is an example (Halpern, 2013). Consider a model whose universe contains nine marbles: five are black and four are white. Let us assume that P assigns a probability of $1/9$ to each marble, which captures the idea that one is equally likely to pick any marble. Suppose the language contains a unary predicate B whose interpretation is the set of black marbles. The sentence $PxB(x) = 5/9$ is true in this model regardless of the assignment.

It is straightforward to check that the counterparts of Kolmogorov's axioms are true in every probabilistic model:

Ax1 $\quad Px(\varphi) \geq 0$

Ax2 $\quad Px(\varphi) + Px(\neg\varphi) = 1$

Ax3 $\quad Px(\varphi) + Px(\psi) \geq Px(\varphi \vee \psi)$

Ax4 $\quad Px(\varphi \wedge \psi) = 0 \rightarrow Px(\varphi) + Px(\psi) = P(\varphi \vee \psi)$.

We notice that if $\varphi(x)$ and $\psi(x)$ are logically equivalent, then for every model $\mathbb{M}$, the set of individuals satisfying φ in $\mathbb{M}$ is the same as the set of individuals satisfying ψ. Thereby $Px\varphi$ and $Px\psi$ must be identical in $\mathbb{M}$. In other words, the counterpart of the Principle of Substitutivity (C9) holds for $Px\varphi$. Given that $Px\varphi$ satisfies $Ax1$ and $Ax2$ which are the counterparts of (C1) and (C2), and it satisfies Ax4 which is the counterpart of (C7), it may be shown on the basis of Theorem 4.40 in Leblanc (1994), that $Px\varphi$ satisfies the counterparts of constraints (C1)-(C6).

As a final observation, we should notice also that when φ is a sentence of the underling first-order language, then Px does not bind any free variable in φ. In that case, either φ is true in the underlying model $\mathbb{M}$ or it is false. In the former case, $\{d \in D : M, g\,[x \mapsto d] \models \varphi\} = D$ and thus $Px(\varphi) = 1$. In the latter case, a similar argument shows that $Px(\varphi) = 0$. In other words, for *sentences* the only values available are 0 or 1.

8.8.3 IF probabilities: some comparisons

Unlike $Px\varphi$, $NE(\varphi, s, \mathbb{M})$ is a concept in the metalanguage. We could lift it to the object language by enriching the syntax with clauses of the form $NE(\varphi) = r$ interpreted by:

- $\mathbb{M}, s \models NE(\varphi) = r$ if and only if $NE(\varphi, s, \mathbb{M}) = r$ iff $\mathbb{M}, s \models_r \varphi$.

Obviously $NE(\varphi)$ behaves like a sentential operator which resembles syntactically $NE(\varphi)$ the degree of belief operator $P(\varphi)$.

On the other side, the interpretation of $NE(\varphi)$, shares with the interpretation of $Px\varphi$ models and model-theoretical semantics. There is, however, one basic difference between the two, apart from the expressive power of the underlying language: in IF logic one considers probability distributions, not over the individuals of an underlying universe, but over the sets of strategies of Eloise and Abelard. Let us look at some of the details.

We have seen already that $NE(\varphi)$ obeys Kolmogorov's axioms of probability.

We now look at the counterparts in IF logic of the constraints (C1)-(C6) keeping in mind Leblanc's observation to the effect that Kolmogorov's axiom (C7), the Principle of Substitutivity (C8) and the constraints (C1)-(C2) pick up the same probability function as (C1)-(C6).

When φ is an ordinary (slash-free) first-order sentence, it follows from (E1) and (E2) that the only probabilistic values taken by φ are 0 and 1. In this case the game-theoretical negation becomes classical, ordinary negation and all the counterparts in IF logic of the constraints (C1)-(C6) are easily shown to be satisfied.

It is the phenomenon of independence which leads to indeterminacy in IF logic and to the violation of the counterpart of Principle of Substitutivity (C8).

The counterpart of this principle in IF logic says that if two IF sentences are logically equivalent, then their probabilistic values are the same in every model. In symbols:

Substitutivity If $\models \varphi \equiv \psi$ then for every model $\mathbb{M}$:

$$M \models NE(\varphi) = NE(\psi)$$

where $\models \varphi \equiv \psi$ is defined as

$$\mathbb{M} \models^{+} \varphi \quad \texttt{iff} \quad \mathbb{M} \models^{+} \psi$$

for every model $\mathbb{M}$.

The Matching Pennies sentence $\varphi_{MP} = \forall x(\exists y/\{y\})x = y$ and the first-order sentence $\psi = \exists y \forall x x = y$ which is logically equivalent to it serve as a counter-example to Substitutivity. We recall that φ fails to be true in any model and so does ψ. (1-element models are excluded in IF logic). Let $\mathbb{M}$ be any model set with two elements. Then, as pointed out earlier:

$$\mathbb{M} \models NE(\varphi_{MP}) = 1/2 \quad \texttt{and} \quad \mathbb{M} \models NE(\psi) = 0.$$

(For the second conjunct we have used the conservativity of equilibrium semantics.)

One could object to the counter-example on the basis of the fact that our notion of logical equivalence is not general enough, for it excludes one element models. It can be shown, however, that even with this broader notion of logical equivalence, substitutivity fails. The Matching Pennies sentence φ_{MP} and the sentence $(\varphi_{IMP} \vee \forall x \forall y x = y)$ will serve as counterexamples, where $\varphi_{IMP} = \forall x(\exists y/\{y\})x \neq y$. In a 1-element model both φ_{MP} and $\forall x \forall y x = y$ are true. Hence $(\varphi_{IMP} \vee \forall x \forall y x = y)$ is also true. On the other side, on any model with strictly more than one element, φ_{IMP} is indeterminate and $\forall x \forall y x = y$ is false, which makes $(\varphi_{IMP} \vee \forall x \forall y x = y)$ indeterminate. We already know that in such a model φ_{MP} is indeterminate too. So the antecedent of the Substitutivity principle is satisfied for the broader notion of logical consequence. Now let $\mathbb{M} = \{a_1, ..., a_4\}$. By P1:

$$\mathbb{M} \models NE(\varphi_{IMP} \vee \forall x \forall y x = y) = max\{NE(\varphi_{IMP}), NE(\forall x \forall y x = y)\} = 3/4.$$

On the other side, we have seen that

$$\mathbb{M} \models NE(\varphi_{MP}) = 1/4.$$

Substitutivity also fails for a stronger notion of logical equivalence, i.e., agreement on both truth and falsity on all models $\mathbb{M}$:

$$\mathbb{M} \models^{+} \varphi \quad \texttt{iff} \quad \mathbb{M} \models^{+} \psi$$
$$\mathbb{M} \models^{-} \varphi \quad \texttt{iff} \quad \mathbb{M} \models^{-} \psi.$$

φ_{MP} and φ_{IMP} will serve as a counterexample also in this case. Recall that both of them are indeterminate on all models (with at least 2-elements). Thus they are trivially strongly logical equivalent. Let $\mathbb{M} = \{a_1, ..., a_4\}$. We have seen that the values of φ_{MP} and φ_{IMP} in $\mathbb{M}$ are $1/4$ and $3/4$ respectively.

It is easy to see that any indeterminate IF sentence φ is a counter-example to both $NE(\varphi \vee \neg\varphi) = 1$ and $NE(\varphi \wedge \neg\varphi) = 0$. For let $\mathbb{M}$ be a model set with two elements. Then

$$\mathbb{M} \models NE(\varphi_{MP} \vee \neg\varphi_{MP}) = max\,\{NE(\varphi_{MP}), NE(\neg\varphi_{MP})\} = 1/2$$

and

$$\mathbb{M} \models NE(\varphi_{MP} \wedge \neg\varphi_{MP}) = min\,\{NE(\varphi_{MP}), NE(\neg\varphi_{MP})\} = 1/2.$$

This is a counter-example to (C2). Let $\mathbb{M}$ a model set with two elements. Then

$$\mathbb{M} \models NE(\neg(\varphi_{MP} \wedge \neg\varphi_{MP})) = 1 - NE(\varphi_{MP} \wedge \neg\varphi_{MP}) = 1/2 \neq 1.$$

It is easy to see that φ_{MP} is also a counter-example to (C3):

$$NE(\varphi) = NE(\varphi \wedge \psi) + NE(\varphi \wedge \neg\psi).$$

Again, on any model $\mathbb{M}$ with two elements we have:

$$\mathbb{M} \models NE(\varphi_{MP} \wedge \varphi_{MP}) = min\,\{NE(\varphi_{MP}), NE(\varphi_{MP})\} = 1/2.$$

We also know that

$$\mathbb{M} \models NE(\varphi_{MP} \wedge \neg\varphi_{MP}) = 1/2.$$

Hence

$$\mathbb{M} \models NE(\varphi_{MP} \wedge \varphi_{MP}) + NE(\varphi_{MP} \wedge \neg\varphi_{MP}) = 1 \neq NE(\varphi_{MP}).$$

In probabilistic semantics the principle $P(\neg\varphi) = 1 - P(\varphi)$ is obtained as a theorem using (C2) and (C3). Although the counterparts of (C2) and (C3) fail in IF logic, the counterpart of the principle $P(\neg\varphi) = 1 - P(\varphi)$ still holds, as witnessed by (P3).

Finally, few words about implication in probabilistic IF logic. We recall that $\varphi \rightarrow \psi$ is defined as $\varphi\neg \vee \psi$.

Both the degrees of belief and the statistical frequency systems satisfy the following principles:

1. $P(\varphi \rightarrow \psi) = 1$ `iff` $P(\varphi \wedge \neg\psi) = 0$
2. $P(\varphi \rightarrow \psi) = 1$ `iff` $P(\varphi) = P(\varphi \wedge \psi)$
3. $P(\varphi \rightarrow \varphi \wedge \varphi) = 1$
4. $P(\varphi \wedge \psi \rightarrow \varphi) = 1$
5. $P(\varphi \wedge \psi \rightarrow \varphi) = 1.$

For a proof of these statements on the basis of (C1-C6), see Leblanc (1994). The proof of all of them relies essentially on (C3). For instance, for the proof of (2) notice that by (C3), $P(\varphi) = P(\varphi \wedge \psi) + P(\varphi \wedge \neg\psi)$. Hence if $P(\varphi) = P(\varphi \wedge \psi)$, then $P(\varphi \wedge \neg\psi) = 0$ and from (1) we get the desired result, $P(\varphi \rightarrow \psi) = 1$.

Although the counterpart of (C3) fails in IF logic, the counterpart of (1) still holds in virtue of (E1) and (E2). But the counterparts of (2)-(5) are not valid, as may be easily seen taking $\varphi = \varphi_{MP}$ and $\psi = \varphi_{IMP}$.

8.8.4 Conclusions

All the three probabilistic systems introduced in this paper satisfy Kolmogorov's axioms for probabilities. If we consider only the first-order subfragment of IF logic, then $P\varphi$,$Px\varphi$ and $NE(\varphi)$ satisfy also the Substitutivity principle (C9) and thereby all the axioms (C6)-(C6). But the full language of IF violates the constraints (C9), and in particular the constraints (C2) and (C3).

The three systems are intended to model different phenomena.

We have already pointed out that $Px\varphi$ is intended to model statistical knowledge as illustrated by the sentence

1. The probability that a randomly chosen brick is black is $^5/_9$

whereas $P\varphi$ is intended to express the degree of belief conception illustrated by

2. The probability that b (a particular brick) is black is $^5/_9$.

It is not quite clear how (2) could be analyzed in Leblanc's probabilistic semantics, apart from simply assigning it the primitive probabilistic value $^5/_9$. Bacchus (1990) and Halpern (1990) propose a more interesting framework in terms of possible worlds and a probability distribution over them. In this new framework, the value of $P\varphi$ is the sum of the probabilities of the possible worlds in which φ is true.

$NE(\varphi)$ arises in the context of the interaction of standard quantifiers in sentence like $\forall x(\exists y/\{x\})x = y$ interpreted in a model set $\mathbb{M}$. We ask the question: what are the odds that when you draw a ball from a set $\mathbb{M}$ and, after you put it back, I will do the same, without seeing the ball you drew, the choices coincide? There is no uncertainty about the world, as in the other two frameworks, but uncertainty about what the other will do. The "odds" in this case are determined by certain rationality assumptions about the behaviour of the players.

References

F. Bacchus, 1990, 'LP, A Logic for representing and reasoning with statistical knowledge', *Computational Intelligence*, 6, pp. 209-231.

F. Barbero, On existential declarations of independence in IF Logic, *Review of Symbolic Logic*, forthcoming, 2013.

F. Barbero and G. Sandu, 2014, Signalling in Independence-Friendly Logic, *Journal of the IGPL,* 22, 638-664.

J. Barwise, 1979, 'On Branching Quantifiers in English', *Journal of Philosophical Logic*, 8, pp. 47-80.

J. K. van Benthem, 2008, 'Games That Make Sense': logic, language, and multi-agent interaction, in K. Apt and R. van Rooij, eds., *New Perspectives on Games and Interaction* (volume 4 of Texts in Logic and Games). Amsterdam University Press, Amsterdam, pags 197-209.

J.K. van Benthem, 2014, *Logic in Games*, MIT Press, Cambridge, Massachusetts.

A. Blass and Y. Gurevich, 1986, 'Henkin quantiers and complete problems', *Annals of Pure and Applied Logic* 32, pp.1-16.

X. Caicedo, F. Duchesne, and T. M. V. Janssen, 2009, Equivalence and Quantifier Rules for Logic with Imperfect Information, *Logic Journal of IGPL* 17, 91–129.

H. Gaifman, 1964, 'Concerning measures of first-order calculi', *Israel J. of Mathematics*, 2, pp. 1-18.

W. Goldfarb, 1979, 'Logic in the Twenties: The Nature of the Quantifier', *Journal of Symbolic Logic* 44, pp 351–68.

T. Grune-Yanoff, A. Lehtinen, 2012, *The Philosophy of Game Theory*, in Handbook of the Philosophy of Economics, ed. Uskali Mäki, Elsevier.

B. Guerrien, 2002, *La Théorie des Jeux*, Economica, Paris.

J. Halpern, 2013, 'Logic and Probability', *Stanford Encyclopedia of Philosophy*, 2013.

J. Halpern, 1990, 'An analysis of first-order logics of probability', *Artificial Intelligence*, 46, 1990, pp. 311–350.

J. C. Harsanyi, 1967, 'Games with Imperfect Information Played by Bayesian Players', *Management Science*, 14.

L. Henkin, 1961, 'Some remarks on infinitely long formulas', in *Infinitistic methods*, Pergamon Press, Oxford-London-New York-Paris, pp.167-183.

J. Hintikka, 1974, 'Quantifiers vs. Quantification Theory', *Linguistic Inquiry*, vol. 5, pp. 153-177.

J. Hintikka, 1996, *The Principles of Mathematics Revisited.* Cambridge, UK: Cambridge University Press.
J. Hintikka and J. Kulas, 1983, *The game of language*, D. Reidel, Dordrecht.
J. Hintikka and J. Kulas, 1985, *Anaphora and definite descriptions: Two applications of game-theoretical semantics*, Dordrecht, Holland.
J. Hintikka and G. Sandu, 'Informational Independence as a Semantic Phenomenon', Logic, Eds. J. E. Fenstad et al. (eds.) *Logic, Methodology and Philosophy of Science*, vol. 8, Amsterdam: Elsevier, 571–89, 1989
J. Hintikka and G. Sandu, 1989, 'Informational independence as a semantical phenomenon', in J.E. Fenstad et al. (eds.), *Logic, Methodology and Philos-ophy of Science* VIII, Elsevier Science.
J. Hintikka and G. Sandu, 1991, *On the methodology of linguistics: A case study*, Basil Blackwell.
J. Hintikka & G. Sandu, 2010, Game-Theoretical Semantics, in: J. van Benthem & A. ter Meulen (eds.), *Handbook of Logic and Language*, Elsevier, Amsterdam, pp. 361-410, 1997
W. Hodges, 1997, 'Compositional semantics for a language of imperfect information', *Journal of the IGPL* 5: 539-563.
A.N. Kolmogorov, 1933, *Grundbegriffe der Wahrscheinlichkeitsrechunung*, Berlin.
H. Leblanc, 1994, 'Alternatives to Standard First-Order Semantics', in *Handbook of Philosophical Logic*, Volume I, D. Gabbay and F. Guenthner (eds.), Dordrecht: Reidel, pp. 189–274.
D. K. Lewis, 1969, *Convention: a philosophical study*, Harvard University Press, Cambridge, 1969.
A. L. Mann, 2011, G. Sandu, and M. Sevenster, *Independence-Friendly Logic: A Game-Theoretic Approach*, Cambridge University Press.
R. May, 1989, 'Interpreting Logical Form', *Linguistics and Philosophy*12, 387-435.
A. Mostowski, 1957, 'On a Generalization of Quantifiers', *Fundamenta Mathematicae* 44, 12-36.
J. von Neumann, 1928 Zur Theorie der Gessellschaftsspiele. *Mathematische Annalen*, 100, 295-32.
M.J. Osborne, *An Introduction to Game Theory*, Oxford University Press, 2003.
M.J. Osborne and A. Rubinstein, 1994, *A Course in Game Theory.* MIT Press, Cambridge, Massachusetts.
K.R. Popper, 1955, 'Two autonomous axiom systems for for the calculus of probabilities', *British Journal for Philosophy of Science*, 6, pp. 51-57, 176, 351.
J. Ratliff, 1997, Lecture Notes, <jim@virtualperfection.com/gametheory>
G. Sandu, 1993, On the Logic of Informational Independence and its Applications, *Journal of Philosophical Logic* 22, 29–60, 1993
G. Sandu, forthcoming, 'Languages with Imperfect Information', in S. Gosh, R. Verbrugge, and J. van Benthem (eds.) *Modeling Strategic Reasoning*, Springer, Texts in Logic and Games, in print.
G, Sandu, 2011, 'Game-Theoretical Semantics', in L. Horsten and R. Pettigrew (eds.) *Handbook of Philosophical Logic*, Continuum Companions to Philosophical Logic, pp 251–300.

G. Sandu, 2012, 'Independence-Friendly Logic: Dependence and Independence of Quantifiers in Logic, *Philosophy Compass*, pp1–21.
M. Sevenster, *Branches of Imperfect Information: Logic, Games, and Computation.* PhD Thesis, Amsterdam: University of Amsterdam, 2006.
M. Sevenster and G. Sandu, 2010, 'Equilibrium semantics of languages of imperfect information', Annals of Pure Appl. Logic 161(5): 618-631.
G. Sher, 'Ways of Branching Quantifiers', 1990, *Linguistics and Philosophy*, 13, 393-422.
B. Skyrms, The Stag Hunt and the Evolution of Social Structure, Cambridge University Press, Cambridge, UK., 2004,
A, Tarski, 1936, 'Wahrscheinlichkeitslehre und mehrwertige Logik, *Erkenntnis*, 5, pp. 174–175.
A. C. Varzi, M. Warglien, 2003, 'The Geometry of Negation', *Journal of Applied Non-Classical Logics* 13:1, 9–19
D. Westerstahl, 1987, 'Branching Generalized Quantifiers and Natural Language', in *Generalized Quantifiers: Linguistic and Logical Approaches*, D. Reidel, Dordrecht.
E. Zermelo, Uber eine Anwendung der Mengenlehre auf die Theorie des Schachspiels, in A.E.H.Love (eds.), *Proceedings of the Fifth International Congress of Mathematicians*, vol. II, Cambridge University Press, Cambridge, pp 5010-504, 1913.